THE QUIET REVOLUTION

Co. Monaghan in the 1940s — the ESB workers are objects of great curiosity.

THE ELECTRIFICATION OF RURAL IRELAND

1946–1976

Michael J Shiel

THE O'BRIEN PRESS
DUBLIN

This edition first published 2003 by The O'Brien Press Ltd,
20 Victoria Road, Dublin 6, Ireland.
Tel: +353 1 4923333; Fax: +353 1 4922777
E-mail: books@obrien.ie
Website: www.obrien.ie
First published in hardback in 1984 by The O'Brien Press Ltd.
Reprinted in hardback and paperback in 2003 by The O'Brien Press Ltd.

ISBN: 0-86278-841-2 (hardback)
ISBN: 0-86278-840-4 (paperback)

British Library Cataloguing-in-Publication Data
A catalogue record for this title is available from the British Library.

2 3 4 5 6 7 8 9 10
03 04 05 06 07

Editing, typesetting, layout and design: The O'Brien Press Ltd
Printing: Betaprint

PICTURE CREDITS

We would like to thank the following for permission to reproduce photographs: The National Library, p.17; GA Duncan, p.189; The Catholic Communications Centre, p.255; Ulster Folk and Transport Museum, p.263; CS Vanston, p.217; Irish Photographic Enterprises, frontispiece; Maurice Curtain, p.217; Ronnie Persse for cartoon, p.133. We also thank the following for permission to quote copyright material: Brandon Book Publishers; *The Irish Times*; Kenny Fine Bindings; *The Irish Press*; Burns and Oates.
Every effort has been made to contact copyright-holders, if any oversight has occurred we ask the holders of such copyright to contact the publisher.

NOTE

Since this book was first published, Ireland has changed currency from the pound to the euro. Prices quoted in the text are in Irish pounds. IR£1 is approximately €1.27.

CONTENTS

TABLES

ACKNOWLEDGEMENTS

IT WOULD HAVE BEEN IMPOSSIBLE to attempt this book without the co-operation of a multitude of former Rural Electrification colleagues. Help was given not only freely but with enthusiasm and delight that a significant chapter in the life of the Irish rural community was to be recorded.

Space would not permit the mentioning by name of all contributors but neither can some names be omitted. Particularly valuable was the contribution of the late William F. Roe who despite failing sight and consequent lack of access to written notes recalled with amazing clarity the early days of the Scheme. Patrick J. Dowling who succeeded Bill Roe as Engineer in Charge was also extremely helpful. Patrick Commins of the Rural Sociology Department of An Foras Talúntais gave me immense assistance both by his advice and by providing research material. Another tower of strength was Patrick Bolger, County Development Officer for Co. Donegal, and author of *The Irish Co-operative Movement, Its History and Development.* Patrick scrutinised my early efforts page by page and his uninhibited but most constructive criticism helped immensely in sharpening up the text.

I must also acknowledge the help of the officials of the various voluntary rural organisations and that of Dr M. T. Casey of Maynooth College for his information on Nicholas Callan. Thanks are also due to former REO colleague Tom Timlin who prepared the groundwork for Appendix 6 (Things Chiefly Technical) with such competence and to Mrs Annie Joye for allowing me access to her unique collection of press cuttings of the early days of 'Rural'.

A very special word of thanks is due to Mrs Kathleen Fitzgerald (née Robinson) who converted my semi-illegible drafts, redrafts and re-redrafts into a most professionally typed final manuscript.

Finally, the story would have been much the poorer but for the rich lode of contemporary information, comment and anecdote available for the mining in the fourteen volumes of the REO NEWS covering the first fourteen years of the Scheme, a publication which rates a chapter to itself in the book.

I would like to dedicate this book to my wife Delma, to thank her for her support and for the countless cups of tea and coffee that kept me going at difficult times.

Michael J. Shiel,
December 1983

INTRODUCTION TO THE 2003 EDITION

Those who established the Electricity Supply Board (ESB) by Government Order on 11 August 1927 ensured that the provisions of the Electricity Supply Act were far-sighted. In addition to being given a mandate to operate, manage and maintain the Shannon Scheme, the ESB was given the general duties of promoting the sale of electricity as well as taking over the control and reinforcement of the supply and distribution function. It is worth recalling the views of Dr Thomas McLaughlin, first Managing Director of ESB. In making the case for the Shannon Scheme he articulated the need for a national network that would incorporate the provision of supply to the whole country:

> 'The people in our remote villages must have the comforts which villagers in other lands enjoy. Electricity, the great key to the economic uplift of the country, must be provided on a national scale, cheap and abundant.'

This book recounts what was to become the second phase of that story, which was undoubtedly one of the most important projects ever undertaken in this country. In telling the story the author highlights the difficulties that had to be overcome, the sense of pride felt by those undertaking the work and the fundamental transformation its successful completion brought to ordinary people's lifestyles. It is wholly appropriate that the reprinting of this book is being organised as one of the final acts of our 75th anniversary celebrations.

It is a remarkable book and it is easy to recognise that it was written by someone with a great understanding and knowledge of the project. It should be no surprise to learn, therefore, that the author himself was one of the key pioneers who contributed to that process. In chronicling this story, Michael J Shiel has performed an invaluable service not only for ESB but for future generations who wish to understand and appreciate what has become known in general parlance as 'The Quiet Revolution'.

Padraig McManus
Chief Executive, ESB

LOOKING TO THE FUTURE FROM THE PAST:

Rural Electrification and Network Renewal

The Rural Electrification Scheme, started over fifty years ago, has been described as the greatest social revolution in Ireland since the Land Reforms of the 1880s and 1890s, which gave farmers the right to purchase their land. A report on Rural Electrification prepared by the ESB at the request of the Government was published by the Oireachtas in 1945, but implementation of the Scheme could not commence until 1946 because of wartime restrictions.

- Activity on the farm and in rural households was dictated by the availability of daylight.
- There was no running water available because water pumps depended on a supply of electricity.
- Heating and cooking depended on solid fuel or timber and turf, which in most cases had to be cut and harvested by the family.
- Cooking was confined to the open hearth or the range.
- It was difficult to maintain hygiene, especially in the dairy, with neither hot or cold running water and without any form of refrigeration.
- Industrial development, even that based on small-scale operations, was not feasible without a supply of electricity.

The Rural Electrification Programme was planned on an area basis and for this purpose the country was divided into 792 areas, priority being given to those areas which yielded the highest revenue in relation to the capital cost of the supply network. The Scheme quickly gathered momentum but there were fluctuations in the pace at which it subsequently proceeded, due mainly to changes in Government policy from time to time. In 1955, for example, the subsidy was withdrawn by the Government and not restored until 1958. In 1962 the Government decided to accelerate the programme by increasing the subsidy from 50% to 75% for a period of five years. In pursuance of this policy the ESB undertook an intensive campaign (including a re-canvass of unconnected houses) to bring supply to as many houses as possible in the period. This scheme, known as Planned Post-Development, was so successful that by 1965 almost 300,000 rural houses (81% of the total) were connected to the supply system at a total cost of approximately €54 million.

- Despite the supposed conservatism of farmers, rural Ireland responded enthusiastically to the opportunity offered for development.
- The advent of electric power utterly transformed rural life in all its aspects – economic, social and cultural.
- The ready availability of power made possible a rapid expansion in both agricultural and industrial output in rural areas.

It can probably be claimed too that without the benefit of a widespread rural supply system, Ireland would have been unable to contemplate membership of the European Economic Community (now European Union). The rapid economic developments of the 1960s had, however, created such a surge in demand for new connections in rural areas that by 1970 it had become evident that the completion date for the Rural Scheme as a whole was a receding target. The ESB took the initiative in formulating new proposals aimed at completing the Scheme by 1975. This third stage was completed in the late 1970s – bringing expenditure on the Scheme to about €140 million – by which time it was estimated that about 98% of all homes in the country had supply.

Since that time many of the electricity networks erected by ESB in the days of Rural Electrification have been replaced and upgraded. In recent years Irish economic performance has been very buoyant and the level of sustained growth achieved has been unprecedented, with the result that during the 1990s the Irish economy was given the title, 'the Celtic Tiger'. ESB is now investing heavily to upgrade the existing network and is actively planning for future requirements well into the next century. The Network Renewal Project is presenting many similar challenges to the development of the original Scheme, but every opportunity will be taken to organise the work so as to minimise supply interruptions for customers and disturbance to farm activities.

The huge investment programme planned over the next five years will probably exceed €3 billion and will be particularly focussed on regions in the midlands and on the western seaboard. This level of investment will be hugely important to the economies and social life of these regions. The Capital Investment Programme is also critical in ensuring ESB can continue to deliver a first-class service to all its customers.

Brendan Delany
ESB Archive & Heritage Manager

Foreword

THIS IS A BOOK OF UNIQUE VALUE, aptly named. It records in considerable detail not only the bringing of light and power to our rural areas, but also the economic, social and cultural revolution brought about by this event. The story told in this book has also, however, a significance beyond the direct effects on the farm and the home and the creation of a more meaningful, rewarding and abundant life for those on the land. It points to many other advantages and lessons arising from this great project.

For generations, the way of life in rural Ireland had changed but little. The same season-by season operations were carried out by human and animal power, often under conditions of great hardship. Farming for many was of a subsistence nature. Life in the rural home had followed the same pattern over many generations. With candle and lamp conditioning living there, many of the housewife's tasks were onerous and time-consuming, sometimes amounting to downright drudgery. Then came rural electrification bringing power into the homes and on to the farms, lessening the burden on the housewife, shortening the time of many chores, providing light and heat at the turn of a simple switch. On the farm it provided the means for much greater efficiency in many operations and a base for the application of modern technology. Again at the turn of a switch the time for such activities as milking, grinding, milling and cleaning could be more than halved, apart from creating greater basic efficiency. It would have been impossible, for instance, to have brought about the great advance in dairying in the last two decades without electrical power.

A means was provided for rural development with power for industries and services which were then badly needed to stem emigration. The forge could be converted into a workshop to serve the needs of farm mechanisation. People from cities and towns could enjoy urban living standards in the countryside. Electricity facilitated social activities in halls and other places where people gathered. In effect, rural society was opened up, parochialism and insularity reduced, and rural people were provided with amenities and opportunities previously the prerogative of urban dwellers.

While all of these advances were not realised at once, the potential was created. Indeed, looked at in retrospect, the changes were penetrating and, as can now be seen, were cumulative. I know that there are some who lamented the passing of the old times and their attributes, but for anyone born and reared in rural Ireland, with experience of day-to-day living there and the hard lives of many, the change has been remarkable and rewarding. As one reads this book, the way in which rural people were provided with new opportunities and the way windows were opened on a much broader spectrum of life emerges clearly.

There were, I believe, many less tangible but nevertheless important benefits from rural electrification which are of relevance and value to national development. High on such a list are the confidence and self-reliance created by the successful prosecution primarily through our own national effort – engineering, planning and management skills – of this mammoth exercise.

As one reads about the details of organisation, of the surmounting of the problems of pole provision, transport and various technical constraints on the innovative input, the picture clearly emerges of an operation carried out by highly-motivated staff, with pride in their work and determination to do a good job. These attributes are badly needed today. Reading this book, I began to realise how privileged I was to know some of the main actors in this scene.

These days we hear much talk and criticism – much of it so ill-founded – of the state-supported or sponsored enterprises. Here, however, we see a project carried through under conditions presenting many constraints and barriers by such a body. It is an example of outstanding enterprise by the public service. There is an important lesson here for the policy-makers and planners. What has to be done should be clearly defined, the required resources identified and decided, and then the organisation concerned should be given a clear go-ahead and allowed to get on with the job. This does not always happen. Fortunately in this case, constraints which could have interfered seriously with this great project were overcome because the organisation and people responsible were determined and highly motivated.

The successful carrying out of this scheme required the involvement of many people other than the staff concerned. Many rural families had to be contacted and convinced: there were reservations against installation on the part of many who would ultimately benefit. The cost factor was a serious consideration for many farming people with limited resources. It was necessary to surmount such barriers and the achievement of progress in this respect, through a system of local committees deeply involving the local community, was a master stroke. It is a fine example of what can be achieved by working positively with people through a participative community approach. Here again there is a lesson to be learned in dealing with many current problems in terms of enriching and effectively utilising the democratic process.

Another related aspect of this co-operative approach is the extent to which the Board effectively mobilised, with beneficial results, resources in other organisations. Again, this is an example well worth following where the discipline and capacity to involve others are such essential ingredients for success.

There is no doubt that this project has achieved much and if more, in terms of farming and rural development, has not resulted it is not because of any shortcoming in its organisation and prosecution. The position in this respect is quite clear. From the agricultural point of view, this scheme provided the base for the application of modern technology on our farms and in the processing plants concerned with agriculture. With as yet only about one-third of our farmers applying modern technology in a productive way, there is obviously much more scope for development. This position can and must be changed by removing the present barriers to progress. In this respect, education, advice, training and demonstrations, so effectively used in this scheme, are pivotal for success.

I feel that it is especially to be regretted that with the first important step taken in rural development through this scheme, the scheme was not further advanced on the basis of a well-planned, creative programme of integrated rural action. This could have ensured the most effective deployment of our resources of land and people. Perhaps some day we will have the vision and courage to proceed along that road. The scheme is a concrete example of what a well-planned input into rural living can do.

Finally, this project was a great challenge to those involved. This story shows that the challenge was met and answered. The Board has rightly ensured that the history and achievements of this project have been recorded for posterity. It is one of the major advances in our country in this century and perhaps in social and societal terms the most significant since the famine, apart from the abolition of landlordism. Michael Shiel has made a penetrating analysis of the scene through his record of fact, sidelight and anecdote, personal and otherwise. He has provided a most interesting and stimulating book for which he and the Board merit deep appreciation and thanks.

Dr Tom Walsh
Former Director of ACOT and of An Foras Talúntais

Introduction

THE STORY OF RURAL ELECTRIFICATION IN IRELAND is the story of political, financial and technical decisions; the procurement of construction materials in a situation of world scarcity; the recruitment, organisation, training and motivation of staff. More importantly, however, it is the story of a rural people left by history and circumstance in a depressed and backward state accentuated by a recent debilitating world war, their initial reaction to innovation, their doubts, the persuasion process and finally the culmination in the acceptance and utilisation of the new energy to achieve a higher living standard and to avail to the maximum of the new market opportunities opening up for their produce.

In telling the story the vital contribution of the various instructional, research and development bodies, the voluntary rural organisations and the improving market situation to the transformation of rural Ireland will be highlighted — a transformation that was not of course entirely due to the advent of rural electrification. Rural electrification was, however, one of the first agents of this change and one of the most essential of the infrastructural developments that made it possible. In those early post-war years it could fairly be said that it was the knocking on the farmer's door by the rural electrification canvasser that first sounded the knell of those old, entrenched, conservative and cautious attitudes, which had kept so many small farmers in the subsistence bracket, their wives condemned to a life of drudgery and so many of their children fated to take passage on the emigrant ship.

Fully to document the human background against which the scheme was carried out would be beyond the scope of this book or, indeed, the capacity of the writer. However, the social, economic and cultural factors which operated at the start and which evolved during the course of the scheme play such an important part that they surface constantly in the telling of the story. Indeed, the promotion of rural electrification could be regarded as a classic case-study in the communication and eventual acceptance of innovation in a largely ultra-conservative society. It is to be hoped that the necessarily brief outlines sketched in the following chapters will prompt some social scientist or historian to probe more deeply into this fruitful field of research.

Free use is made throughout of quotations from contemporary official and newspaper reports, magazine articles, speeches and letters. We are fortunate in having had scribes who at the time not only recorded the facts but who captured the flavour, as a modern writer could not, of rural Ireland immediately after the war and the reactions of its less sophisticated society to the first rumblings of this quiet but momentous revolution which was so completely to change its way of life.

The first rural consumer — putting away the old oil lamp in McCullagh's public house, Oldtown, 15 January 1947.

Prologue

'Somebody – I cannot remember who – switched on the lights . . .'

THE PLACE WAS OLDTOWN in north County Dublin. The date was 15 January 1947, a cold, windy winter's evening with patches of snow. The body of the small village hall was packed with local people, while up on the stage sat a group of leading citizens, the parish priest and senior ESB officials. The occasion was the switching on for the first time of electricity under the new Rural Electrification Scheme. There was, however, a snag of which the audience was unaware. Severe winter storms had caused a last-minute fault in the supply line and even now, as the speeches commenced, a line crew was working frantically in the pitch darkness over a mile away to put things right.

At the back of the stage was mounted a large switch, which, when operated, should illuminate the hall and village with the new light. The hands of the clock now showed eight, the scheduled time for the 'switch on'. As the Engineer-in-Charge, W. F. Roe, commenced his speech he kept one eye on a small table at the side. There sat a gramophone turntable, connected to the still inanimate supply line. Anxiety sharpened as talk time was running out. Suddenly and unobtrusively the turntable commenced to rotate; the pick-up dropped onto the record, and legend has it that a very relieved Bill Roe concluded his speech to the strains of 'Cockles and Mussels, Alive, Alive — O'!

A blessing was invoked. The switch was thrown. The hall burst into light and Oldtown passed into the history books as the first village in Ireland to be electrified under the Rural Electrification Scheme.

At the end of the year, R. M. Smylie ('Quidnunc'), discussing the memorable events of 1947 in *The Irish Times,* closed his column as follows:

> But how many of these things will be remembered in, say, 2047? I dare swear that if any event is recorded in the history books (taught through the medium of Russian) it will be none of those I have mentioned; rather it will be one which has passed almost unnoticed, amid the turmoil of the year.
>
> Somebody – I cannot remember who – switched on the lights in some village — I cannot remember where – and rural electrification took her bow. And if that does not mean more to the country than all the rest of the year's events put together, I shall be very surprised indeed.

CHAPTER ONE

Electricity for the Nation

IT COULD BE FAIRLY CLAIMED that one of the fathers of rural electrification in Ireland in the twentieth century was a man born in an Irish rural community in the closing days of the eighteenth century.

Nicholas Joseph Callan was born at Darver near Dundalk, Co. Louth, on 20 December 1799. He went on to study for the priesthood at St Patrick's College, Maynooth. At college his intense interest in the pursuit of scientific knowledge brought him far beyond the traditional curriculum. His genius and devotion to scientific discovery eventually led to his appointment as Professor of Natural Philosophy at St Patrick's College and to a series of basic discoveries and inventions, particularly in the field of electromagnetics. He achieved considerable international recognition, but far less than was warranted by the fundamental importance of his work.[1]

Callan's invention of the induction coil in 1836 and his pioneering work contributed greatly to the development and evolution of the power transformer without which the widespread distribution of electricity would have been well-nigh impossible, especially in sparsely populated rural areas. His output of ideas and inventions was prolific. In 1837 he made a fundamental discovery, the principle of self-excitation in dynamo-electric machines. His communication on this is dated Maynooth, 20 February 1838 — twenty-eight years before Werner von Siemens reported his observations on the same principle to the Berlin Academy of Science. The Encyclopaedia Britannica of 1860 estimated the lifting capacity of Callan's giant electromagnet of 1834 – which, happily, may still be seen in St Patrick's College museum – at over one tonne.

It would perhaps be too much to expect that Callan, labouring away in the tranquility of his Maynooth laboratory, set as it was in an Ireland where millions of depressed poverty-stricken peasant farmers were struggling for mere existence, could ever have envisaged that a century later the fruits of his labours would be employed to help provide a better life for their descendants. The journey from the induction coils of Maynooth in the 1830s to the distribution transformers of rural Ireland in the 1940s was indeed to be a long one, but the experiments of this dedicated professor, himself of rural stock, may fairly be regarded as the first vital steps on that journey.

Following the general pattern of the times, the practical development and utilisation of electricity in Ireland was first confined to the cities and larger towns. Callan was credited with causing some amusement among his colleagues in the 1830s by predicting the use of electricity for lighting. He would have been vindicated when in 1880 the electric filament lamp was invented almost simultaneously by Edison and Swan. In that year the Dublin Electric Light Co. was set up and in the same year an experimental arc-lamp was erected outside the offices of the *Freeman's Journal* in Prince's St. in Dublin. The following year saw seventeen public lights – arc-lamps – in the vicinity of Kildare Street, Dawson Street and St Stephen's Green in Dublin. The first provincial town in Ireland to have public electric lighting was Carlow, supplied from a generator in a flour mill some four miles away. The year was 1891. Charles Stewart Parnell, who was addressing a large meeting there on the night of the 'switch-on', used the new light as a symbol of a new and free Ireland in his speech.

Slowly at first, and then more rapidly, electricity spread to all the principal towns in Ireland, supplied in some cases by the local authority and in others by privately owned supply companies. By 1925 there were 161 separate electricity undertakings in the country. These operated on many different standards and voltages and almost all were DC systems. This type of separate development, beneficial as it was to the cities and towns concerned, could not form any basis for the general extension of electricity to the rural community. This necessarily had to await the establishment of a national alternating current electricity grid with adequate generating capacity. It was in order to provide this nation-wide electricity supply system that the concept of the Shannon Scheme was developed in the early years of the newly created Irish Free State.

CHAPTER 2

The Shannon Scheme

THE SHANNON SCHEME was the brainchild of a young Irish engineer, Thomas A. McLaughlin, who left Ireland in December 1922 to work with the German firm of Siemens Schuckert. In his work on the supply of electricity to various parts of Germany, he became convinced that electricity was the key to the economic uplift needed by Ireland. At this time the Irish Free State, in the aftermath of the Great War, the War of Independence and the Civil War, was in a very depressed condition. McLaughlin carried out a large amount of preliminary work on his own before approaching Siemens or the Irish government, but when he did it was with a very carefully thought-out scheme of utilising the waters of the River Shannon to generate cheap electricity for the whole country. The core of his scheme was the harnessing of the fall of the river, not at a number of separate points as had been mooted hitherto, but utilising almost the whole fall at one point, Ardnacrusha.

In concentrating on the Shannon, McLaughlin was flying in the face of a strong Dublin lobby, which was recommending that the Liffey should be the first river to be developed in this fashion, as the big demand for electricity would come from the Greater Dublin region. The extension of supply to outside the metropolitan area did not appear to be a high priority.

On the other hand, the McLaughlin/Siemens report clearly envisaged the inclusion of Irish farms in its proposed scheme.

> Ireland is, no doubt, as were all the other countries, faced with the problem of lessening the dullness and hardship of the farmer's life and, in addition, with the problem of labour on the farm. The remedy abroad is electricity and no doubt in Ireland it will serve the same end. It brings to the farmer first of all the most pleasant and rational form of lighting, serves to brighten up his home and, by lighting his yard, barns and cow-houses, makes it much easier and more agreeable for him to work there on a dark night or an early winter's morning. Secondly, it provides for the most efficient

The daily drudgery of getting water from the river in early twentieth century Ireland.

> application of labour, as the electrical drive is the only possible economic mechanical drive for farm machinery, especially on a small farm.[1]

After many discussions the project, which had now been developed in detail in conjunction with Siemens, was accepted by the government of the new Irish Free State. There was a lot of criticism from politicians, business interests and newspapers, Irish and British. The *Irish Independent* and *The Irish Times* when not actually hostile expressed doubts about the size and timing of the venture while the *Morning Post* of London said, 'The present needs of Southern Ireland cannot be more than about 50 million units per annum, whereas the scheme provides for 150 million units. The Irish people . . . . with such an excess of power . . . . may all be electrocuted in their beds.' (Apropos of this and many other similar prognostications made at the time, it is worth recording that this figure of 150 million units was reached in 1935 and that by 1970 annual consumption of electricity *from rural consumers alone* exceeded 1,000 million units).

Patrick McGilligan, Minister for Industry and Commerce, was enthusiastic from the start and undertook the extremely difficult task of piloting the Shannon Scheme through the political storm. On 13 August 1925 the contract for the scheme, which was to cost £5.2 million, was signed between the government and Siemens Schuckert. The time for completion was three and a half years.[2] The official opening was on 22 July 1929, and the first current commenced to flow in October of that year.

The Electricity Supply Board was established on 11 August 1927, with the objective of operating, managing and maintaining the Shannon Scheme and distributing and selling its output on a national scale. It also got the task, which the government regarded as being of key importance, of promoting and encouraging the purchase and use of electricity and of controlling, co-ordinating and improving its supply, distribution and sale.

In setting up the first state-sponsored business organisation to manage the nation's electricity supply, the government of the day showed considerable foresight. Banking and business interests and even the Farmers' Party were vocal in advocating private enterprise. To many the concept of a state-run undertaking was anathema. One newspaper described it as 'the first fruits of Bolshevism in this country'.[3] While the government did for a while examine the possibilities of involving the private sector, there were major difficulties.

In the first place, large sums would have to be raised to finance the huge capital expenditure involved in the widespread extension of the

Top Left: Dr Thomas McLaughlin. *Top Right*: Patrick McGilligan. *Above*: Callan's induction coil, which can be seen in Maynooth College museum.

national electricity network. In addition, considerable sums would be required as compensation in the takeover of existing undertakings. The return on such investment would of its very nature be extremely long term. In the perilous economic conditions following the recent birth of the State, such long-term capital was unlikely to be available from Irish investors. The possibility of attracting foreign capital was investigated and dismissed as it was obvious that foreign investors would only be attracted on their own hard-nosed commercial terms which would not harmonise with the social and economic objectives of the scheme. Furthermore, the idea of delivering control of such a vital national resource to a foreign corporation was in those early days of independence completely unacceptable. As investigations proceeded, it became obvious that nothing short of a publicly owned organisation could successfully overcome all the obstacles in the way of achieving the ultimate object of the Shannon Scheme, that of providing a cheap, reliable, integrated electricity service nation-wide.

An obvious option was the setting up of a separate government department. It was considered, however, that the detailed accountability of a government department would not afford the flexibility and freedom of decision required. What the government was looking for was an organisation which would be answerable to the Oireachtas and which would implement national policy as directed, but which would have maximum freedom to deploy its resources, exercise its business judgement and make its own commercial decisions. The result of the search was the creation of a state-sponsored or 'semi-state' Corporation which was to be a blue-print for many other such bodies in years to come.

The Electricity (Supply) Act 1927, setting up the ESB gave the Board powers to sell electricity either in bulk to other distributors or directly to the consumers, but shortly after its constitution the Board made decisions which were to have a profound effect on the pace of development.

It would not sell in bulk to other distributors of electricity. It would retail directly to the customer on a 'non-profit-making' basis (i.e. after making suitable provision for interest, sinking fund, depreciation and other such charges).

It would acquire all existing electricity undertakings.

It would develop its own technical expertise in the design, construction, and operation of electricity systems.

These decisions formed the basis for the creation of a single integrated and country-wide electricity supply system and for the development of native 'know-how', which were of immense value when the time came in getting rural electrification off the ground.

The decision not to sell in bulk even to the large, experienced public-supply authorities such as Dublin Corporation and Rathmines and Pembroke Townships was opposed violently. The ESB, however, backed by the government, regarded the proper development of a national electricity supply as something that should not be subject to municipal boundaries or local politics. Time has shown that this decision, contentious as it was at the time, was undoubtedly the right one. It permitted a single-minded national approach to electricity development, untrammelled by the necessity to deal with and harmonise the interests of a multitude of local supply authorities, as is still the case in a number of western European countries.

It was also decided to decentralise all customer-oriented activities, including power-line and substation construction. This was done by setting up sub-organisations on a geographical basis – 'Districts' – and giving these the requisite resources and the maximum autonomy in providing a full electricity service. The boundaries of these Districts and the smaller 'Areas' into which they were subdivided were determined by the needs of the system and of consumers. They did not necessarily have to follow county or urban boundaries.

PROMOTIONAL STRATEGIES

In setting up the ESB, the government was fully aware that the physical construction of a supply system would not in itself be sufficient. The approach of the average householder to electricity was, it could be said, extremely circumspect. The price per unit charged by existing suppliers was in the region of one shilling (5p; equivalent to about £1 in 1983 terms). The ESB was able to offer electricity at about 1p per unit throughout the 1930s. However, price alone was not enough to ensure the best use of the new 'Shannon' current. The consumer needed to be educated, advised and persuaded to make the Shannon Scheme a paying proposition. This was regarded by the new Board as the most important of its tasks. A number of strategies were developed.

The ESB defined its *pricing policies* as being promotional for growth 'to encourage consumers to forsake their conservative habits'.

A *sales organisation* with a contract and wiring department was established with the object of reducing the risk of defective wiring and ensuring safe operation.

Outlets for the *sale of tested electrical appliances* were opened in all the main cities and towns.

The Board did not consider its task fulfilled with the arrival of the supply at the consumer's meter. It also set up a *service repair organisation* for virtually all types of electrical apparatus to ensure that appliances were repaired and returned to service in as short a time as possible and at an acceptable cost.

These promotional efforts were effective. By the outbreak of World War II 170,000 consumers had been connected and were using 320 million units per annum, almost double the output of the Shannon Scheme. The bulk of the remainder was generated in the Dublin coal-fired station at the Pigeon House. This achievement was confined almost entirely to the cities and towns but again, as with many other activities, the expertise developed was of great value when the time came to make the same promotional effort among the rural community. Even during the war years some expansion was achieved, so that by 1946 the number of consumers had reached 240,000, using 380 million units per annum. Again, however, the great majority were 'urban' consumers. The 400,000 rural dwellings had been virtually untouched.

CHAPTER THREE

Rural Electrification – Overtures

THE ULTIMATE OBJECT OF THE SHANNON SCHEME – supply on a *nation-wide* basis, rural as well as urban – had not been forgotten. As early as 1925, Professor Boyle of University College, Cork, wrote an article in the journal *Studies* on 'The Possibilities of Electricity in Agriculture'. Against the background of the proposed scheme, he developed in some detail the economic advantage of using electricity in various farmyard tasks such as grain-crushing, root-pulping, chaff-cutting, milking, milk-separating and churning. He finished his article by pointing out that 'If the farmer can be shown that the utilisation of electricity saves him time and money and makes life more tolerable, the rural demand is likely to surpass the experts' expectations'.

In April 1927 Seán McEntee, himself an engineer and later to become Minister for Finance in the first Fianna Fáil government in 1932, read a paper on rural electrification before the Irish Centre of the Institution of Electrical Engineers. The minutes of the centre record that, owing to the importance of the paper and of the subject dealt with, it was decided to postpone discussion to a special meeting to be held a fortnight later.

A perusal of the minutes of the Irish Centre of the Institution around this period discloses an increasingly high degree of interest by electrical engineers in rural electrification. In March 1928 a paper was read by R. Borlase Matthews on 'Electric Ploughing'. Discussion broadened into the general question of electricity in agriculture. In December 1931 two British members, E. W. Dickinson and H. W. Grimmet, came to Dublin specially to present a paper on 'The Design of a Distribution System in a Rural Area' and in February 1938 J. S. Pickles read a paper on 'Rural Electrification'. Thomas McLaughlin was elected Chairman of the Irish Centre of the Institution in 1940 and, as was to be expected, his opening address was mainly on the subject of rural electrification. In March 1942 J. C. Costello presented a very detailed treatise to the centre on the economics, suggested technical design and utilisation of a rural electrification scheme. In February 1944 a further paper entitled 'Some Aspects of Rural Electrification' was presented by J. F. O'D. McFaul and J. M. F. Higgins.

McLaughlin never forgot that the completion of the Shannon Scheme and the extension of 'Shannon' electricity to the cities and towns of Ireland was only part of the task. Many many times during the thirties he raised the matter of rural electrification at board meetings (he had been appointed an executive director and member of the board by the government). In April 1936, at his instigation, a schedule of rates of charge for supply to villages of not more than 250 population and isolated consumers in rural areas was produced. This provided for a 'fixed charge' based on the floor area of the dwelling-house and, at a lower rate, of the farm out-offices. In addition there was a kilowatt-hour (kWh) or 'unit' charge of 1.25*d.* per unit for general domestic purposes (1*d.* per unit for cooking and ½*d.* per unit for water heating). The fixed charge for farm premises, including out-offices, worked out at about 75% more than the corresponding charge in urban locations (which was based on Poor Law Valuation of the premises). For the average small farm, it came to 30*s.* (£1.50) per two-month period. This in itself was not unreasonable; the snag lay in the high cost of making the initial connection.

The Board required that total expected revenue, taken over two years, had to be equal to or greater than the capital cost of connection. This requirement was generally referred to as the '2 to 1 ratio'. Any balance in the capital cost not so covered had to be contributed by the householder. While this was not too onerous where houses were clustered together, requiring no capital contribution in most towns and villages, it had the effect of excluding practically all isolated rural premises. There the capital costs involved would be so high that there would be no possibility of generating the required level of revenue.

In 1937 McLaughlin proposed that the ratio should be extended to 4 to 1 from the existing 2 to 1 level, pointing out that a ratio of 5 to 1 was quite common in other countries. The Board, however, postponed consideration of the proposal until the question of the large capital involved in a comprehensive rural electrification scheme was definitely settled. It should be noted that even the extension of the ratio as proposed would achieve a very modest advance in rural electricity development, as supply to most rural houses would still require very high capital contributions from the householders. Examples given by McLaughlin for premises half a kilometre to two kilometres from an existing ten thousand volt (10kV) line quoted contributions of from £78 to £280.

This assessment of contributions was on the basis of a strict accounting approach to the costs involved and again there is a record of McLaughlin pleading for a mitigation of this and the adoption of marginal costing. However, it was obvious to the Board that at best any mitigation it could apply would only benefit those householders within comparatively easy reach of existing electricity networks. Only a very small proportion of rural dwellings lay within even the two kilometre range. The capital contribution for the remainder – the vast majority – would be so high as to be unthinkable.

It should be stressed that the widespread penetration of electricity into rural areas depended on the prior development of the ESB transmission and distribution system. A backbone of transmission and distribution lines had first to be built to carry electricity to the cities and towns. This was the task on which the Board concentrated in the thirties and early forties. It involved very heavy capital expenditure (about £18 million by 1946), which was advanced by the government on a totally repayable basis at around 5% interest (the ordinary bank rate at the time). In order to meet its interest and repayments and at the same time hold down the price of electricity, the ESB had to fix strict financial conditions to the connection of new consumers.

These arrangements, which had been worked out for urban development would be totally unsuitable for rural conditions. A drastically new approach to organisation, planning and financing was needed. McLaughlin was fully aware of this and his constant raising of rural electrification at Board meetings was undoubtedly intended to stir the Board and ultimately the government to action.

The government itself was under increasing political pressure to do something for the farmers. Seán Lemass, Minister for Industry and Commerce, was an ardent advocate of rural electrification as a means of improving the lot of the rural dweller. In May 1939 he requested the ESB to prepare plans to supply rural areas with electricity and to make proposals regarding finance etc. The ESB immediately undertook a detailed investigation. This was directed by Dr McLaughlin who was assisted by Patrick J. Dowling and Alphonsus J. McManus. As well as being qualified engineers, these two men were farmers' sons with a good appreciation of the problems of the rural dweller. They came from opposite ends of the country: Paddy Dowling grew up on a large farm in County Carlow, while Alfie McManus's family farmed in the Lagan valley area of east Donegal. By September 1939 good progress had been made but the outbreak of World War II brought other priorities to the fore.

In the early years of the war the ESB, with a very restricted staff, was preoccupied with keeping electricity supply available to the community under very difficult conditions. It could not see its way to devote much time or resources to planning rural electrification for what then appeared to be a distant and uncertain future. Nevertheless, the government, beset though it was with grim wartime problems, still regarded post-war rural electrification as one of its priorities. Lemass subsequently recalled (Dáil Éireann, 24 January 1945):

> In 1939 when I ceased to be Minister for Industry and Commerce [he had been appointed Minister for Supplies] and the present Minister for Local Government was taking over from me I drew his attention to certain matters which were outstanding. One of the matters outstanding was this report from the ESB on rural electrification. During his period as Minister for Industry and Commerce, Mr McEntee pursued the Board – almost to the point of friction – to produce this report.

Lemass, while remaining Minister for Supplies, took back the Industry and Commerce portfolio in August 1941. In the autumn of 1942 he wrote to ask the ESB formally if its plans for rural electrification were ready. The reaction in the ESB was one of astonishment. 1942 had been a very difficult year and the mere maintaining of electricity supply had taxed the resources – and resourcefulness – of the Board to their limits. An acute drought in the early part of the year had severely restricted the output of Ardnacrusha and had thrown most of the load on to the coal-burning Pigeon House station. This depended on irregular supplies of very poor British coal – pithead rejections, in fact, as neutral Éire was very far down on the coal priority list – which was extremely difficult to burn. A severe electricity shortage threatened. New connections of domestic electricity consumers were discontinued, those for industrial purposes closely vetted and restrictions introduced for all existing consumers. The German armies were at the gates of Stalingrad and the prospects for rural electrification could not have appeared more remote.

Nevertheless, it was obvious that Lemass meant business. McLaughlin and his team set to the work with such effect that by 22 December 1942 a comprehensive report had been completed and delivered to the Department of Industry and Commerce. The report, which was subsequently published in a slightly edited and revised form as a White Paper, is outlined in some detail in Appendix 1. At this stage it will suffice to give the main features, which were as follows: an analysis of the problem of rural electrification in Ireland; a review of its progress in other countries; an outline of the proposed method of development on an 'Area' basis;

Seán Lemass, to whom Irish industry and agriculture owe a great debt.

a detailed treatment of the costs and the possibility of recouping these from revenue (about 12% of the capital cost per annum would be required); the conclusion that a capital subsidy would be necessary and that in order to keep costs and subsidy down to a reasonable level, supply under the scheme could only be offered to 86% of rural premises. Of these it was expected that only about 80% would elect to take supply so that the scheme as put to the government provided for supply to 280,000 premises, i.e. 69% of the estimated 403,000 dwellings in rural Ireland. The estimated cost of supplying these was £14 million at pre-war prices.

The government, undoubtedly pressed on by Lemass, wasted little time in considering the report. On 26 August 1943 the ESB received a letter from the Department of Industry and Commerce approving the proposals subject to a number of stipulations and giving certain government undertakings, which may be summarised as follows.

Current 'rural' tariff rates to apply (i.e. the 1936 schedule), subject to any war increases generally applicable.

Electricity to be supplied to all rural premises where the capital cost of connection did not exceed sixteen times the prospective revenue from the fixed annual part of the tariff.[1]

The State to compensate the ESB for any shortfall in the required 12% return on the capital expenditure as yielded by the revenue from the fixed annual part of the tariff (based on 5% interest rates).

The ESB to plan at once to commence construction in the maximum number of centres simultaneously (not less than one in each county) as soon as supplies of material became available.

Subject to above, priority to be given to the most remunerative areas.

Legislation to be prepared forthwith to enable the scheme to proceed.

The Board to plan to complete the scheme within ten years of the date on which adequate supplies became available.

The maximum amount of equipment to be of Irish manufacture.

On 28 August a statement was issued by the Government Information Bureau to the effect that work on the scheme would commence as soon as materials became available after the war. The report was published in the form of a very detailed and well-illustrated White Paper of 114 pages in August of the following year, 1944.

On 29 November 1944 the Minister for Industry and Commerce, Seán Lemass, introduced the Electricity (Supply) (Amendment) Bill 1944, a massive document covering hydro-electric schemes, acquisition and management of fisheries (by the ESB), particular powers and duties of the Board and advances to the Board out of the central fund. The section (number 41) covering the Rural Electrification Scheme, which was to be one of the largest projects undertaken by the ESB, was surprisingly curt, comprising less than half a page – four short paragraphs – in a document of seventy-seven pages. However, it was sufficient to launch the scheme. The section provided for an initial advance of £5 million out of the central fund, of which a half would be repaid to the fund out of monies provided by the Oireachtas and half by the ESB. The Bill, when passed, became the 1945 Act of similar title.

In practice, it was agreed that the ESB repayments would be phased over a period of fifty years for which purpose a sinking fund was set up. The Act provided for the repayment of the other half of the advances (the subsidy) to the central fund out of voted monies, the purpose being to ensure that the initial borrowings to cover the subsidy moiety of the advances would be repaid out of revenue.[2]

Even though the section dealing with rural electrification occupied but a small portion of the total Bill, the debate on it was an extended one with many long contributions by spokesmen for all parties. These included Patrick McGilligan, by now in opposition, who had in earlier years piloted through the Acts setting up the Shannon Scheme and the ESB. He recalled some of the pressures exerted on him to drop the Shannon Scheme: '... we were prayed by businessmen, by chambers of commerce and by every newspaper in the country to stop this thing and to cut the losses, to pay the German firm something to get out of the whole business'. There was some criticism of certain aspects such as the fixed charge, but in general there was almost complete approval and a strong awareness of the significance of rural electrification in improving the lot of rural dwellers. The contribution of Deputy James Larkin Junior in quite a short speech towards the end of the debate was typical in demonstrating this awareness particularly in the area of social change:

> In our country, electrification is more than merely producing light and power; it is bringing light into darkness. The great value of this proposed scheme of rural electrification is not that we are going to have farming by electricity instead of by hand, but that we are going to put into the homes of our people in rural areas a light which will light up their minds as well as their homes. If we do that, we will have brought a new atmosphere and

> a new outlook to many of these people. It often appals me to think that we have just under 2,000,000 men and women, the majority of whom are in many physical ways, debarred from mental development, the application of their mental faculties, just because they have not got the elementary physical help of a decent light with which to read. If we can get them light and nothing else, then I think we have brought about a great change.

In the Senate debate of 7 March 1945 the same awareness of the benefits of the proposed scheme was manifest on all sides of the chamber.

In his closing speech in the Senate the Minister emphasised the role which electrification could play in easing the lot of the rural housewife:

> I hope to see the day that when a girl gets a proposal from a farmer she will enquire not so much about the number of cows but rather concerning the electrical appliances she will require before she gives her consent including not merely electric light but a water heater, an electric clothes boiler, a vacuum cleaner and even a refrigerator.

CHAPTER FOUR

Paying the Piper

THE WHITE PAPER (see Appendix 1) had made it clear that the primary problems to be solved in floating the Rural Electrification Scheme were financial ones. The very low income levels of most rural dwellers meant that rates of charge which would be adequate to service the very high capital investment would be beyond the means of the majority of householders to pay. Some form of subsidy was necessary. In the heavily industrialised countries such as Britain and Germany the proportion of rural dwellers was small and this subsidy could be provided by urban electricity consumers in the form of a small and tolerable increase in their electricity charges. In Ireland, however, in 1946, there were only about 250,000 urban electricity consumers. To expect these to subsidise the electrification of over 400,000 isolated rural dwellings, or even the 280,000 mentioned in the White Paper, was, in the ESB's book, completely out of the question. In the initial years of the scheme this principle appeared to be accepted by the government of the day, but later governments took a different approach.

Throughout the thirty-odd years of the scheme the question of the amount of subsidy required and who would provide it remained a constant issue between the ESB and the government. In simple terms it could be said that the ESB was prepared and even anxious to extend supply to every premises, however remote, provided it received sufficient government subsidy to prevent the scheme becoming too big a burden on ESB funds. The government, on the other hand, had to meet many other calls upon the public purse, which restricted the amount available for rural subsidy. Nevertheless, as time went on there was intense political pressure on the government to extend supply deeper and deeper into the uneconomic areas. Naturally much of this pressure was transferred to the ESB and we have the story of various government approaches, ESB reactions, subsequent agreements, disagreements, compromises and, on occasions, government directives. The end result was that a scheme which started out in 1946 to supply 69% of the rural premises in the State over a ten-year period aided only by a government subsidy finished thirty years later with 98% of all premises connected, but with the help of a large cross-subsidy from the urban electricity consumers.

Those thirty years saw many ebbs and flows in the economic tide, in capital availability, in interest rates and in the fortunes of the farming community. Nevertheless, once started, the work of rural electrification went on without a break. At some times more monetary assistance was available from the central government than at others. At some times the ESB was chided by the government for going too slowly; at others, warned of the unavailability of capital to meet all its proposals. At times political pressures to go beyond what was provided for in the statutes, or what was possible from available resources, were intense; at others it looked as if the completion of the scheme was a receding target as more new houses continued to be built in rural areas while demands for improvement and strengthening of supply to existing consumers cut deeply into the resources available. Rising capital costs were always a problem. The original estimate based on pre-war prices for connecting 280,000 consumers was just over £14 million (an average of £50 per consumer). By 1946, when the first pole was erected, costs had escalated by 50%, and by the end of the initial phase in 1962 the average capital cost of connecting a new consumer had risen to £150. Each rise in capital costs, which was not necessarily accompanied by a corresponding rise in fixed charges, had ramifications in the economics of the scheme. The ratio of fixed charge revenue to capital cost, on which the 'coverage' (i.e. the percentage of householders to whom supply could be offered at standard rates of charge) was based, was frequently upset. (See Appendix 3 for a fuller treatment of this.)

THE INITIAL SCHEME

In 1946 when a start appeared to be possible capital costs had increased by 50% on pre-war estimates, while the increase allowed by the Minister in fixed charges was only 20%. This decision upset from the very outset the basis of the economic calculations. It was now assessed that a capital subsidy of 89% would be required to give the originally envisaged 'coverage' (i.e. supply to 280,000 consumers without capital contribution). The ESB at this stage continued to insist that rural electrification as an enterprise would have to stand on its own feet with the help of whatever subsidy was available. There was to be no question of a shortfall in the rural account being subsidised internally from surpluses in the

Their armour is nothing but oilcoats and boots
Their weapons are climbers and pliers
Their battles are fought up where high tension shoots
And danger's the song of the wires.
Anon., 'REO News', December 1950

non-rural accounts. Rising capital costs would thus require a higher subsidy, so long as the rural fixed charges were not allowed to rise in the same ratio. At the beginning, this also appeared to be the approach of government, but in later years a change in the economic climate and in government outlook compelled the ESB to alter its stance.

In 1946, however, this was still for the future. In these initial stages the government, while acknowledging the ESB's case, was reluctant to increase the subsidy of 50% incorporated in the 1945 Act without actual experience of events. In even agreeing to a fixed subsidy at all it was going against the advice of the Department of Finance. That department, mindful no doubt of earlier problems in the controlling of ESB expenditure, had counselled against any fixed rate of subsidy. In a memorandum dated 2 March 1944 it had drawn attention to the uncertainty of the degree of subsidisation needed. This issue arose because the return required by the terms of the White Paper (12% per annum) and the annual fixed charge revenue provided for were merely estimates. The memorandum suggested that as an alternative to a fixed rate of subsidy, the Minister for Finance would 'whenever he thought proper' decide on the amount of subsidy necessary to enable the Board to break even. It expressed the belief that in the early years with the 'best' areas being developed first, no subsidy at all might be required.

As could be expected, the ESB did not relish the implications of this suggestion. If accepted, every aspect of the scheme which affected financial results – rates of charge, construction costs, working costs, overheads, etc. – would be subject to detailed investigation by the Department of Finance. A system of dual management would, in fact, be set up. The ESB advised the government that if this were to be so the Board could not be expected to carry responsibility for the success of the scheme. On the other hand, a fixed subsidy would throw the whole responsibility for sound management squarely on to the ESB. The government took the point. It accepted the principle of fixed subsidy and thus removed the threat of detailed financial control by the Department of Finance. In his speech in the Senate debate on 7 March 1945 Seán Lemass stressed his trust in the ESB's own control: 'We can, I think, with confidence entrust them with the task of preparing these plans and carrying them through without any of the safeguards which were considered necessary in the early stage of the Board's existence.'

Nevertheless, the government was still reluctant to agree to an increase in the rate of subsidy above the 50% incorporated in the 1945 Act. In July 1946 it instructed the ESB to go ahead with the scheme on the basis of a 20% increase in (pre-war) fixed charge rates and said that 'should the actual expenditure on the scheme disclose justification for a subsidy in excess of the 50% provided for in the Electricity (Supply) (Amendment) Act 1945, the additional subsidy will be paid out of the Transition Development Fund for a period of two years and will subsequently be provided by means of amending legislation'. In the same letter, the ESB was asked to proceed with the erection of the network with all possible speed.

The Department of Finance had indeed been justified in highlighting the tenuous nature of the financial projections in the White Paper. The economics of the scheme were very finely balanced and, because of the lack of time and resources to carry out more widespread field work (the Report was prepared in the height of the war), were based mainly on desk research. There was no certainty about how things would work out on the ground. The fact that selection of the order of area development was on the basis of 'the best annual fixed charge as a percentage of capital' might tend to bring the areas with the best return to the fore in the early years of development, giving a high initial return and perhaps a misleading impression of the eventual out-turn. As the poorer and more remote areas were reached, the return on the capital might drop drastically. On the other hand, increasing familiarity with the progress and benefits of the scheme might tend to give a higher number of 'acceptances' in the areas developed later, and thereby a higher return so that the best areas might not necessarily be those developed first. No one could be sure at this stage and the decision of the government to keep its powder dry was probably correct.

Some control on expenditure was established by providing in the Act for only £5 million i.e. one quarter of the total estimated cost, thereby ensuring that the progress of the scheme came under Dáil scrutiny at any early date. From the ESB's point of view, this control was of course vastly preferable to the close, continuous scrutiny of its expenditure by the Department of Finance which had earlier loomed up as a possibility.

EARLY SNAGS

By the start of the scheme agreement had been reached between the government and the ESB to offer supply free of capital contribution to all householders in a selected area where the capital cost involved did not exceed 17½ times the annual fixed charge revenue. This represented a return on capital of only 5.7% — the extreme case. Other, more favourably situated customers in the area would give a better return. In fact the White Paper had estimated an average return country-wide of 9.7% based on pre-war costs, but the 50% post-war increase in capital costs as against the allowed 20% increase in fixed charges had brought this figure down to 7.76%.

But what of the premises returning less than the minimum of 5.7%? According to the terms of the scheme, houses where the return was less than this *even marginally* were excluded *completely* from the benefits of the scheme. They would by definition fall outside the 86% of all premises for which the scheme was intended to cater. It quickly became obvious that such a sharp cut-off point was unacceptable. Many would-be electricity customers were being informed that they did not qualify for inclusion as their 'return' fell below the requisite 5.7%, sometimes by only a small amount. To the unfortunate householders thus excluded, particularly in the marginal cases, this appeared as a stupid bureaucratic quibble and it was not long before irresistible pressure developed for a change in the system.

THE SPECIAL SERVICE CHARGE

In December 1947, just a year after commencement, the demand for change was met by the introduction of the special service charge (SSC). This provided for an annual supplementary payment by householders whose premises fell outside the required minimum figure. The special charge was such that when added to the 'normal' fixed charge (according to the schedule) it gave an overall return on the capital cost of 7.76% (the average national return provided for). The supplementary payment would thus be small for houses just outside the limit, growing progressively larger for the more remote premises.

At the time this was accepted as a satisfactory approach to the problem. Later these special service charges were to become the subject of much criticism and controversy. Many rural Dáil deputies and county councillors understandably could not see the justice of a practice which, in their eyes, penalised those of their constituents who lived in worthy townlands but by chance found themselves at a distance from the power-line, which meandered erratically through the countryside, 'designed no doubt by an intoxicated engineer', as one aggrieved householder put it!

SUBSIDY

Throughout the whole of the scheme the question of adequate subsidy was constantly an issue between the ESB and the government. The accounting approach of the ESB with regard to capital expenditure on the rural scheme was determined by the fact that any capital costs not met by non-repayable grants would have to be met from borrowings. For the first ten years these monies were provided by the government's central fund and thereafter by loans floated on the national and international money markets. Provision would have to be made out of revenue for the payment of interest on such loans and for their eventual redemption. It was agreed with the government that a fifty-year period would be allowed for redemption of central fund loans and an appropriate sinking fund was set up. In addition, a depreciation fund (based on an average life of 22½ years) was created to enable replacement of assets to take place. As in the case of the sinking fund this also required appropriate annual payments from revenue. (This double provision was to become a controversial item in later years.) Finally, there were a number of 'other' costs (operation, maintenance and repair of the network, general administration etc.) which also had to be met from revenue. It was estimated on the basis of experience with urban networks that an annual figure of 4% of capital expenditure would be required to meet these.

In the early years interest rates varied from 2½% to over 5%. On the basis of an interest rate of 3¼% which obtained from 1947 to 1952 the annual charges would amount to 11.15% of the capital cost of the network, thus:

	% of capital cost
Interest	3.25%
Sinking fund (50 years)	0.82%
Depreciation (22½ years)	3.08%
Other costs (operation, maintenance, repair, administration etc.)	4.00%
Total	11.15%

It is worth stressing that once a network is erected these charges must be met irrespective of the actual quantity of electricity used. They thus form the basis of the fixed charge portion of the electricity tariff.

Under these circumstances the effect of a capital subsidy was to relieve the interest and sinking fund charges. Thus a 50% subsidy would reduce these two charges by 50% or in the above example by 2.03% of capital costs. It is also worth noting that a 100% capital subsidy (i.e. a free grant of the complete cost) would still leave annual charges of 7.08% of capital to be met by the fixed charges. An area returning less than this on its capital cost would therefore involve the ESB in a continuous annual loss even if the total cost was met by a free grant. This issue was to rear its head constantly as the scheme progressed and it was found that rapidly rising costs reduced the returns from many areas to levels far below this minimum.

In the early months of 1949, however, when the scheme was just over two years old (about thirty areas had been completed and eleven thousand customers connected) it was hoped that the average return from areas developed in future years would be around 8.5%. This would require a capital subsidy of 65% to break even.[1] By this stage materials were more freely available and the ESB was in a position to step up the rate of construction. Before committing itself to this higher capital expenditure the Board wished to settle the question of adequate subsidy.

It requested that the 50% rate be increased to the 65% figure. A request was also made for the implementation of the undertaking, given in 1946, that the government would top up the subsidy shortfall on the first two years' working. However, a new government was now in power and, following the devaluation of sterling in 1949 and the outbreak of the Korean war in 1950, was experiencing great difficulty in providing adequately for capital investment. It was therefore reluctant to implement the undertaking given by the previous government. The Board was invited by the Taoiseach, J. A. Costello, to put its case to the cabinet committee on capital expenditure of which he was chairman. On 20 October 1950 a deputation from the ESB met the committee. The government's difficulties were outlined and, as a result, the ESB agreed to continue on the basis of a government undertaking to review the subsidy at the end of 1952, i.e. at the end of a five-year in lieu of the original two-year period.

SEPARATE RURAL ACCOUNTS

Around the end of 1949 the government requested that the figures relating to the rural scheme be extracted from the Board's main accounts and balance sheet and shown separately in the annual accounts. This would give it some basis for monitoring the out-turn experienced and assessing the Board's case for a higher subsidy. The request was not received with any great enthusiasm by the Board. Naturally it was keeping separate costings so as to control and monitor expenditure on the scheme, but, as much of the ESB system was utilised to supply both rural and non-rural customers, many of the figures used were necessarily based on estimates and apportionments. To quote the Board's Chief Accountant, 'financial accounts are required to be statements of fact. It has been proven adequately that exact facts in regard to the economy of any location of supply, or any type of supply or any rate of charge are quite unascertainable.' The ESB offered to produce not accounts as such but an 'informative statement'. This was not acceptable, however. A further request for separate accounts drew the joint response from Chief Engineer, Chief Accountant, and Engineer-in-Charge Rural Electrification (6 January 1951) that 'the Board should be reluctant to commit itself to the production or publication of accounts purporting to show true financial results and any figures supplied to State Departments should be given as statistical statements and with all the proper reservations'.

In June 1951 the government changed and Seán Lemass was once more Minister for Industry and Commerce. He in turn continued to press for accounts, but conceded that specific explanations and apportionments could be shown where required. He emphasised that without this information he would find it difficult to assess the ESB case for extra subsidy. The Board was checkmated, and in October the basis and presentation of a rural revenue account was agreed and an account for 1950/51 sent to the Minister. This distributed the total debits to the production account between non-rural and rural revenue accounts. The rural revenue account was charged on the basis of the estimated bulk supply of units and of the estimated rural proportion of the total system peak demand, the first being used for the distribution of fuel and operating costs and the second for capital and administrative charges. Other expenses such as maintenance and repair were based on direct working allocations. On this basis, a surplus of £22,700 was shown. There was, however, a particular point made by the Board with regard to the depreciation charges which, if agreed, would have reduced this figure to almost zero. In the account depreciation was charged, according to standard practice, on the opening capital investment. In view of the heavy and rapidly expanding rate of rural construction there was a strong case for basing depreciation instead on the average capital expenditure for the year. This would result in an extra depreciation charge of the order of £20,000 which would virtually absorb the surplus.

MINIMUM RETURN REDUCED TO 5%

By the end of 1951 costs had risen as a result of devaluation and price increases. These increases had the result of pushing a large number of premises hitherto returning the required 5.7% or better below this figure and made them liable for special service charges. In order to relieve the position the ESB decided on two important changes. Firstly, it dropped the minimum 5.7% requirement to 5%, thereby restoring the proportion of premises to which supply could be offered without service charge.[2]

Secondly, it eased the situation for the remaining premises so that the SSC was now only that required to bring the return up to the new minimum of 5% instead of the 7.76% hitherto demanded. In the absence of an increase in the subsidy or an upward revision of the fixed charges both these concessions had of course the effect of depressing the overall economic return of the scheme.

SCOPE OF SUBSIDY EXTENDED

In July 1952 the government, at the request of the ESB, gave two very important rulings on the interpretation of the Act of 1945. It ruled that capital spent in bringing supply to premises in a rural area after completion of the initial development would qualify for subsidy. This ruling was to become very important in the sixties and seventies when a very large number of premises were connected under various 'post development' schemes.

The government also ruled that the cost of reinforcing the rural distribution system ('system improvements') subsequent to initial development qualified for subsidy. This ruling was of the greatest importance in the development of the scheme and in the optimum use of available capital at any time. It meant that in the early stages the ESB could save on capital investment, or, more correctly, use the available capital to connect a larger number of consumers. It did this by taking advantage of the fact that in the initial period after connection the average new electricity consumer's demand tended to be low and generally took some years to reach the level of established consumers. The requirements of some grew very rapidly while those of other consumers showed very little growth even after many years. It was not possible to predict accurately how a particular consumer's demand would develop, depending as it did on many factors. What the engineer could do, however, was to install the minimum capacity to meet the consumer's likely short-term requirements, but design the network so that it could be reinforced at minimum cost as and when the demand grew. The Board could thus ensure that the minimum amount of capital investment was lying idle at any time in the knowledge that when it subsequently reinforced the system, it would not have forfeited the government subsidy involved in the extra work. On the other hand, if it had been ruled that subsidy applied only to initial development, prudence and the interests of the consumer would have compelled the Board to provide at the outset for the long-term potential demand at a much higher initial – and overall – cost.

CHANGE IN BASIS FOR SELECTION

The White Paper had stressed that rational development required that areas closest to the existing 10kV system should be given priority in development. This was in fact substantially achieved by selecting those areas giving the highest financial return. As with many other aspects of the scheme, however, social considerations and resulting political pressures challenged the rational approach. Areas with a high sign-up frequently found themselves pushed down in the queue by areas with lower acceptance rates because of the more favourable location of the latter. Seán Lemass was particularly concerned that Gaeltacht areas, mostly in remote locations, should not find themselves at the end of a long waiting list and he continually pressed the ESB to revise its selection procedure. In July 1953 a new selection system was agreed whereby the percentage acceptance figure in an area up for selection carried equal weight to the return on capital, thereby giving those remote areas with a high sign-up an opportunity to move up in the queue.

Costs continued to increase. Interest rates, which in the initial stages had been 2½%-3¾%, had in 1953 reached 5¼%. With ever-rising material, labour and money costs and a declining average return per area the gap between annual fixed charge revenue and the annual costs which the subsidy was intended to meet was widening. There was, however, no slackening in the pace of development. In fact, at the insistence of the government the pace was stepped up and by November the rate of connections was 50% up on that of the previous year. For the year ending 31 March 1954, sixty areas had been developed and 23,477 consumers connected bringing the total to 289 areas completed and 101,000 consumers connected which was over one-third of the target.

By February 1954 capital costs were 50% higher than at the commencement of the scheme in 1946 but fixed charges had been pegged at the 1946 level. Most of the areas being selected at this time were returning only about 5% and in a detailed report to the Board on the finances of the scheme, the Engineer-in-Charge noted that 'we have not yet got into what might be called the "lean territory" '. There was a probability, he went on, that as the ESB moved into these areas the rural scheme would built up a large revenue loss. If this happened the Board would have to consider the questions of altering the rates of charge or carrying some or all of the deficit on the rest of the ESB's business. It was not long before this prediction was fulfilled.

WITHDRAWAL OF SUBSIDY

In June 1954 an inter-party government took over once more in a situation of great national financial difficulty. A fresh look was taken at the overall working of the ESB and at the necessity for the rural subsidy. The government's attention focussed on the overall surpluses earned by the ESB in the years following the introduction of rural electrification. As shown in Table 1 these were substantial.

TABLE 1

ESB Trading Results 1946 to 1954 as per Annual Reports

Year	*Surplus*	*Deficit*
1946/47	£248,940	—
47/48	£ 74,996	—
48/49	£212,158	—
49/50	£239,344	—
50/51	£514,107	—
51/52	£ 962	—
52/53	—	£488,213
53/54	£233,209	—

The 1946/47 and 1947/48 surpluses eliminated an accumulated deficiency in the net revenue account. The Board then decided that any future surpluses should go towards building up a contingencies reserve. By 31 March 1951 this reserve stood at £800,000 but was drawn upon after the deficit in 1952/53 of £488,213. This deficit was attributed to an abnormal drought, which reduced the hydro generation at Ardnacrusha by 30% and increased fuel costs by almost £½ million. The rural electrification account which first appeared for 1950/51 showed a surplus of £22,700 and for the next three years showed deficits of £46,000, £10,000 and £30,000 respectively, minute amounts in the context of the total ESB trading.

At the end of October the ESB received a letter from the government noting the substantial net surpluses achieved in every post-war year except in the abnormally dry year of 1952/53. It appeared (to the government) that the Board's liability to repay 50% of the capital cost of rural electrification including the payment of interest had had but little effect on the profitability of its operations as a whole. In fact in the previous year the Board could apparently have undertaken the full liability for rural electrification and still have shown a clear net surplus of over £100,000. The Minister, the letter went on, would be glad to learn if the Board would agree that it should be possible without detriment to the rate of future development of the scheme, to terminate the provision of subsidy by the State, say from 1 January 1955. This was not all. It was also proposed that the Board would 'assume liability to the Central Fund in respect of the moiety of advances for rural electrification repayable from voted monies', which totalled £2.1 million at 31 March 1954. In addition the Board would become responsible for the repayment to the central fund of the full amount of advances for rural electrification in the calendar year 1954 together with interest on each such advance from the date of the advance to 31 December 1954. These last two proposals meant not alone would the subsidy be withdrawn for the future but the Board would have to repay about £4 million already advanced plus accrued interest. (In the event the actual amount came to £4.8 million.)

The Board's reaction as conveyed to the Minister was one of 'surprise and anxiety'. (This was, to say the least, expressing it mildly!) The Board emphasised that the question of subsidy had been examined exhaustively, and agreed with the government of the day, before the rural development scheme commenced. The scheme had been undertaken as a matter of urgent national policy on a positive assurance that a capital subsidy of not less than 50% would be given and the Board had committed itself very heavily as a result. The question of subsidising the scheme from non-rural activities had never been raised. On the contrary, it had been taken as fundamental by the Board that the rural activities would not be permitted to affect adversely the electricity costs of the Board's urban consumers who numbered 250,000 at the commencement of the scheme.

A carefully documented case demonstrated that many of the arguments used by the Minister were not, in fact, based on the true situation or did not coincide with the Board's projections. It dealt in detail with the necessity for building up adequate reserves and pointed out that the current rapid growth of capital charges on new and costly installations could lead to future revenue deficits and that materials and labour costs were still rising, as indeed were fuel prices. The ESB also pointed out the risk that higher rates of charge to rural consumers to compensate for the loss of subsidy might well produce a situation of diminishing returns in that fewer and fewer consumers would get connected, which could possibly lead to a collapse of the scheme. Finally, it stressed that so long as rural electrification continued, capital still had to be provided and provided in the year the development took place. Savings to the community would not result from the withdrawal of subsidy; the effect would be merely to change the incidence of the cost (i.e. from the taxpayer to the urban electricity consumer). With the prospect of heavy fresh charges to be borne, it would have to re-consider its whole policy on rural electrification.

The Minister in turn expressed his feeling that the Board's views were unduly pessimistic. He considered, having regard particularly to the results of the Board's operations in the year ended 31 March 1954 that the undertaking had reached the stage when it was in a position to bear all proper interest and depreciation charges and provide adequate reserves for contingencies without any further subvention from voted monies.

Finally, on 8 March 1955, the debate was terminated abruptly when the Minister formally advised the Board that the government had decided that the subsidy should be discontinued and the ESB would be required to meet the full charges except on capital already repaid to the central fund from voted monies.

Thus, the capital subsidy, the case for which was so carefully developed and documented in the White Paper and which was intended to meet the extra 'social' costs of rural electrification, was withdrawn. From this time on, even though in a couple of years the subsidy was to be restored, it never again met the original objective. The 'loss' on rural electrification increased steadily, as costs increased and as the minimum return on capital was held far below an economic level, from £30,000 in 1952/53 to £3.2 million in 1970/71, the last year in which separate rural accounts appeared in the ESB annual report. By then the cumulative 'loss' recorded amounted to over £19 million. In effect, that figure represented the amount of cross-subsidy from non-rural consumers.[3]

THE PEAK IS PASSED

By 1956 the ESB could foresee a rapidly increasing financial deficit on its rural activities. In August of that year there was a severe tightening of the State capital budget and the ESB was requested to reduce its capital demands on the central fund. As might be expected the Board considered, among other actions, reducing the pace of the rural scheme, which for 1955/56 had accounted for over £4 million or 12% of the State's capital expenditure. It intimated as much to the government and indicated that it was examining the possible cancellation of orders for materials and a reduction in the number of rural construction crews from forty to ten.

Very promptly the Board was advised that the Minister did not agree with the proposed reduction; nor did he agree that any action should be taken to cancel orders for materials. Any reduction in the expenditure of the Board that might be necessary because of the current financial difficulties should be achieved otherwise than at the expense of rural electrification. Nevertheless, reduction was effected as capital expenditure dropped from £4.16 million in 1955/56 to £3.7 million in the following year and to £2.5 million in 1957/58 with only about half the number of new connections.

The reason for the phasing down of the construction rate was not altogether financial. By the end of 1956 over two-thirds of the 280,000 connections envisaged in the scheme had been achieved and it was considered necessary to ease down in a controlled manner so as to avoid a sudden stop. This would bring serious redundancy problems, not alone in the Board's own labour force, but in all the native industries that had grown up to supply materials for the scheme. (In 1957 69% of all material used, amounting to £1,442,000, was provided by Irish suppliers.)

THE SPECIAL SERVICE CHARGE IS QUESTIONED

Almost as sacrosanct in the eyes of the ESB as the capital subsidy was the principle of the special service charge. This allowed householders who otherwise would find themselves excluded from the scheme, to be included on paying a special additional annual charge. In August 1956 the government enquired of the Board as to the possibility of abolishing the special service charge. A long reply from the Board outlined the history of the charge and its purpose. The number actually paying these charges was comparatively small – about 6,000 – and their abolition in the case of this number was not of any consequence. What was of consequence was that if the Board was compelled to supply all applicants without a special service charge, the most isolated premises could demand a supply. In that eventuality it might be faced with connecting about 40,000 very uneconomic consumers at an extra capital cost of £7 million (involving an increase in the annual loss on rural electrification of about £500,000). It put these points very forcefully to the Minister and the issue was dropped for the time being.

THE 'POST DEVELOPMENT' CONSUMERS

In the autumn of 1956, in view of the projected loss in the rural revenue account for 1956/57 (close on £½ million), the growing activity in connecting rural 'post development' consumers now came under particular scrutiny. These were householders who had been left out in the initial development but who now were requesting connection. Even though the main construction crews had moved on, it had been the practice to accede to their requests wherever possible but at a somewhat higher minimum return (now 6.6% as against 4.6% for first-time development). The Board now decided that in view of the scarcity of capital and the withdrawal of the subsidy, it must devote most of its available resources to the programme of initial development of areas to which it was committed. A householder in a developed area who had already been offered and had refused supply and who now wished to be connected must contribute a larger proportion of the cost. A minimum return of 9% was now fixed for all future 'post development' consumers. The Board also abolished the facility of special service charges for these consumers. If the normal fixed charge on a premises did not realise the 9% return, a capital contribution was required from the householder to make up the difference.

The decision was to provoke a strong reaction from the applicants, their public representatives and the farming organisations. Seán Lemass was not long back in his old office, with the change in government in March 1957, before a deputation from the National Farmers' Association raised the issue. He in turn exhorted the ESB to mitigate the conditions and restore the special service charge for the consumers in question. The ESB took a hard line: it pointed out that with the subsidy abolished it was in no position to yield such financial concessions, as to do so would merely accelerate the plunge into the severe loss-making situation which was forecast.

In a comprehensive survey of the financial position of the rural scheme, the Board set out its predictions: by the 31 March 1958 £23 million would have been spent on rural electrification with a £2 million subsidy from the government (i.e. 9% instead of the 50% subsidy which was provided for in the original enabling Act). The effect on the rural revenue account would be as follows.

Retrospective withdrawal of £4.8m subsidy	Loss of £245,000 p.a.
Withholding of £4.7m subsidy earned from date of withdrawal to 31 March 1958	Loss of £300,000 p.a.
At 31 March 1958 annual deficit on rural electrification would be	£650,000 p.a.
At this date 184 areas would remain to be done costing £7m and the estimated deficit on this work would be	£550,000 p.a.
The ESB would thus have spent £30m (subsidy £2m) and would have incurred as a result an annual deficit of	£1,200,000 p.a.

In addition the cost over the years of connecting all remaining premises under 'post development' would amount to a further £6 million.

Once again as in the case of its selection policy the 'rational' approach of the ESB had to yield to political considerations. Seán Lemass would not accept no for an answer. By the following February the capital contribution from post development consumers was abolished; the required minimum return was reduced from 9% to 6.6% and the special service charge restored where the return fell below this.

THE SUBSIDY RESTORED

The case made by the ESB, even though it was a loser in this skirmish, had obviously made some impression on the government. In December 1958 the Electricity (Supply) (Amendment) Act 1958 restored the subsidy of 50% but only for capital expenditure incurred from 1 April 1958. The shortfall of £9.5 million up to this date was never recovered.

CHAPTER FIVE

Building the Organisation

SEÁN LEMASS in his introduction of the 1945 Electricity Bill in the Dáil on 24 January had stressed that a job of the magnitude of the proposed Rural Electrification Scheme had never before been undertaken. It would use over one million poles and would involve the construction of 75,000 miles of new line (as against the total of about 2,000 miles which then existed), the erection of 100,000 extra distribution transformers (as against the current 1,200) and the connection of 280,000 new consumers. Some doubts were voiced whether this huge task could possibly be completed before the end of the century: in the Dáil debate Deputy Daniel Morrissey said that he had heard a period of seventy to eighty years mentioned. The Minister, however, assured him that the aim was to connect 280,000 consumers within ten years of materials becoming generally available.

This presented a big challenge to the ESB engineering organisation. Let us take one item, line construction: at its peak the rural programme would require the construction of over 6,000 miles of line per year involving the erection of 100,000 poles in forty different localities. The fastest rate reached hitherto, at the height of the Shannon Scheme work, was some 650 miles of line per annum. Procurement of the necessary materials would likewise involve quantities far exceeding previous experience, this with difficult post-war shortages of all classes of materials.

The problems had been studied by the Chief Engineers of the Design, Operations and Consumers Departments in the ESB, who identified two important steps that would have to be taken to get the scheme off the ground. The first was the setting up of a separate organisation concentrating on rural construction and development only, under the direction of one leader of proven ability who would have senior (even Director) status and who should be given a very high degree of autonomy and authority covering design and construction, procurement of materials

Above: The work-horse of the rural scheme, the Fordson 10 van.
Below: Starting out 1951 — John Murphy, with bicycle, and Seamus Hayes.

and recruitment and training of staff. The second was a level of decentralisation of the work itself far exceeding previous practice, which would mean passing down to the field engineers as much responsibility and authority as possible.

In October 1944 the Board commissioned P. J. Dowling, assistant to Thomas McLaughlin, to prepare a report containing proposals on the organisation of the scheme. He was requested to have it available in one month.

One of the fruits of the early Board decision in 1929 to set up the District organisation was now manifest. There existed an effective decentralised framework onto which could be grafted the rural construction and development structure required. There were twelve Districts in all, but two of these – Dublin City and Cork City would not be involved in rural electrification. The other ten Districts between them covered all rural Ireland. These 'rural' Districts were Athlone, Cork No. 2, Dublin No. 2, Dundalk, Galway, Limerick, Portlaoise, Sligo, Tralee and Waterford. The District administrative and operational headquarters was in the city or town after which it took its name and the geographical boundaries of the Districts were matched to the electrical 'feeding' and operational layout of the network so that supply to every section and every consumer within the boundaries was under the control of the staff of that District headed up by the District Engineer. For this reason, ESB and county boundaries did not always coincide.

When the main Shannon Scheme contractors departed in the early thirties the ESB had to rely on its own resources to continue the extension of the supply network to the towns and villages of Ireland. This it did by developing skilled crews who worked out of the various District headquarters. When the big expansion was called for by the Rural Electrification Scheme, these headquarter crews often provided the nucleus of skilled and experienced supervisors, electricians and linesmen around whom the new rural construction crews were formed.

A further asset of the District framework was that it provided a readymade decentralised structure for the administrative and financial control of the work. Thus, a very important component of the desired framework already existed which had proved itself both efficient and effective. The task now was to build on this and work out in some detail how the total package of responsibilities and various levels of authority should be divided between the Districts and Head Office.

The District Engineers were able to draw on their experience in building up the urban network to assess the probable impact of the rural work and the level of authority and responsibility that could be taken on at District level. P. J. Dowling involved them in intense consultation and by the end of the stipulated month had produced a report embodying a comprehensive set of outline proposals on the organisation of the scheme in the field. This was accepted in principle by the Board. The report emphasised that, because of the magnitude of the work and its dispersed nature, and because it involved such detailed dealings with so many people, as much administrative work as possible should be carried out locally, and as little as possible in Head Office.

A suggested break up of functions was as follows.

Local Functions:
Selection of areas on basis of rules devised by Head Office
Canvass and development
Layout of networks
Issue of wayleave notices and hearing of objections
Planning of construction
Requisition, storage and transport of materials
Erection of networks and connection of consumers
Local publicity and promotional schemes
Control of staff
All associated clerical and accountancy work including credit approval.

Head Office Functions:
Overall development of organisation and general control of staff
Network design standards and construction methods
Procurement of materials, tools and transport
Determination of rates of charge
Formulation of rules for selection of areas
Development of applications of electricity to agriculture
National publicity and assistance in local publicity
Monitoring of progress and financial results.

The report recommended the setting up of a separate rural organisation in each District. It considered that by doing so it would be possible 'to develop a single-minded enthusiasm for the work which otherwise would be impossible to achieve'. It was felt that in the absence of this enthusiasm and concentration of effort on the one big job, rural electrification, nothing like the desired rate of progress or financial results could be obtained. In order to ensure co-operation and harmony between the existing District 'urban' personnel and the new 'rural' group, and to utilise fully the experience available among the former, it was proposed that the District Engineer should be placed in charge of the rural electrification work in each District while still retaining control of the urban work.

The size and composition of the geographical unit which would form the rural area was then discussed with the conclusion that the parish, with an average area of 25–30 square miles and containing about five hundred premises, would be a suitable unit. Proposals were made on various details of organisation in the selected areas and at District Office level, most of which were incorporated into the final structure.

Dealing with Head Office, the report referred to the wealth of experience and expertise which was already available in the various departments. It suggested methods of utilising these resources without impinging on the autonomy and responsibility of the Rural Office. It discussed the degree of autonomy which should be given to the Rural Office and its relationships with the Board, Head Office Departments, and the Districts.

Finally, the report summarised its recommendations as follows.

That the parish should be the geographical unit.

That the work would be carried out by a temporary office set up in each 'selected' parish.

That these area offices should be subject to a separate Rural Electrification Office in each District Headquarters.

That the District Engineer should be placed in charge of this office while remaining in charge of the existing District Office dealing with urban supply and the transmission system.

That a separate Rural Electrification Office should be set up in Head Office.

That this office should correlate and be responsible for all rural electrification activities in Head Office and the Districts.

That the Board should state what authority it wished to delegate to this office.

That the Board should decide if a Director should be in charge of this office and if so, that it should nominate that Director. Alternatively, should the Board not adopt this course, it should nominate someone else to take charge and define the relationship of the Rural Office to the Board.

The Board agreed to the recommendations in principle but decided against putting a Director in overall charge of the scheme. It invited applications from senior engineering staff for the position. William Francis Roe, District Engineer, Cork City, was selected and appointed on 19 January 1945.

W. F. Roe, a native of Kilkenny city, graduated in 1925 at the age of twenty-one with first class honours in mechanical and electrical engineering. After short periods with the Dublin Corporation Electricity Department and the Shannon Board of Control (set up in 1925 by the government to monitor the progress of the contractors to the Shannon Scheme), he joined the newly formed Electricity Supply Board in 1928. Here he specialised in the conversion of existing town electricity networks to ESB standards and he found himself working closely with a contemporary – P. J. Dowling – who was responsible for building the new distribution networks in towns that had not hitherto had electricity supply. It was the commencement of a long association between the two men, who thus found themselves architects of the ESB distribution system from its conception.

With the setting up of the ESB's District Organisation in 1929, W. F. Roe embarked on a fifteen-year period of service as District Engineer, in Portlaoise, Waterford and finally in Cork City. Wherever his job brought him he involved himself in community affairs, and in 1938 he met Fr John (later Canon) Hayes, founder of Muintir na Tíre. This was the commencement of a strong friendship and of a deep involvement in a fast-growing movement whose aim was community development in rural Ireland. He developed a profound knowledge of the social and economic problems of the countryside and a strong commitment to their amelioration.

Roe was a man of strong personality and a decisive style of management. He had an extraordinary – and to some, disconcerting – ability to express complicated issues in simple 'countrified' language. He had a natural flair for leadership and a demonstrated commitment to the improvement of the rural social and economic scene. These qualities were now allied to extensive experience of the business of electricity distribution. His first task was to recruit the nucleus of senior staff who would assist him in the huge undertaking that lay ahead. His most pressing need was for a high-calibre deputy and assistant in the overall organisation, planning and supervision and he was very pleased when Patrick J. Dowling was appointed.

P. J. Dowling, the farmer's son from County Carlow, was also an honours graduate in electrical and mechanical engineering. On joining the ESB he worked on the building of urban electricity distribution networks for the new Shannon Scheme. This was followed by periods in Athlone, Dublin City and Head Office, mostly working on consumers' problems. His command of engineering economics and his analytical ability involved him in various investigations with Dr McLaughlin and as already recounted he played a major part in the production of the original report on rural electrification subsequently to be issued as the White Paper. In 1944, at the request of the Board, he had prepared a comprehensive set of organisational proposals for the Rural Electrification Scheme, most of which were subsequently adopted. In doing this he had consulted all the District Engineers and had obtained many useful suggestions, including a very thoughtful and valuable contribution from the then District Engineer, Cork City, W. F. Roe now to be his immediate chief. The Roe–Dowling partnership formed in 1945 was to survive through a quarter of a century until both men retired almost simultaneously in 1969.

The third member of the original rural electrification triumvirate was an accountant. Cornelius A. (Neil) O'Donoghue was seconded by the ESB's Chief Accountant to look after the specific accounting problems of the scheme and to ensure their harmonisation with the Board's general accounting system. Memories of grave accounting problems in the fledgling years of the ESB were still green. The whole emphasis had been on getting the construction work done, at the expense of proper accounting control. The new Rural Scheme would also involve an intense construction programme with correspondingly high capital expenditure. It was determined by all concerned that there would be no repetition of the accounting chaos of the early thirties and that strict accounting procedures would be introduced to operate from the very start of the scheme.

Neil O'Donoghue had transferred to the Board from the private electricity sector, moving from the local undertaking in his native Bandon to be manager of the Dungarvan Electric Light and Power Company prior to its absorption by the ESB. He had qualifications not only in accountancy but in electrical engineering, having achieved graduate membership of the Institution of Electrical Engineers and also membership of the Chartered Institute of Secretaries. His talents and training were thus not confined solely to accountancy matters. He knew the technical as well as the business origins of every figure in his books. He played an important part in getting the scheme off the ground in the critical early years and his untimely death from a heart attack less than two and a half years later, was a grievous loss to the rural electrification organisation.

Annie Joye was the fourth member of the Rural Electrification Office staff. A competent woman of strong personality, she acted as secretary

Above: The 'wiring-gang'. *Below*: Landing the electricity cable from the mainland to supply Valentia Island.

to the Engineer-in-Charge. As the organisation grew she built up an efficient corps of REO typists, which she supervised until her departure thirteen years later to take up a promotional position in the personnel branch of the ESB. During her period at the REO every new recruit was inducted willy nilly into becoming a subscriber to Mrs Joye's Mater Hospital (Belfast) Football Pools!

On 14 March 1945, the three founder members of the new rural electrification staff held their first meeting to develop a detailed plan for carrying out the Scheme. There was no formal agenda. The trio had available to them two documents, the ESB report on Rural Electrification (the White Paper) and the report and recommendations on organisation prepared by P. J. Dowling. W. F. Roe recalled:

> We knew that the time programme was ten years, that over one million poles and other items in equally astronomical numbers must be procured and erected. The greater part of the world was still at war. We did not know when the war would end or what the post-war world could be like. Perhaps it was just as well. The Board had appointed us as the nucleus of an organisation to carry out the rural scheme. The organisation's functions were to construct the networks, connect the consumers and see that the job was carried out as economically as possible.[1]

No time was wasted and on 27 April Roe presented to the Board a detailed set of proposals for the organisation of the Rural Office. The document of fifty-two pages reflected the importance of establishing at the very outset a clear understanding and agreement with the Board, on objectives, responsibilities, levels of authority and relationships. It made clear and detailed proposals about what authority should be retained by the Board and what should be given to the Engineer in Charge. It also dealt with areas where he in turn could pass certain levels of his authority further down the line. The main issues dealt with were:

General arrangements for the Organisation of Rural Electrification
Responsibilities of the Rural Electrification Office
Activities and staff of the Rural Electrification Office
Rural activities and staff in District Offices
Activities and staff of Area Offices
Staff Control
Manufacture of materials
Relationship of the Rural Electrification Office to the Board and to the various Departments of the Board
Recruitment and training of staff
Stages of expansion.

The final pages of the document listed steps which should be taken immediately so that the Scheme could swing into action as soon as materials became available. These included such items as the clear definition of *rural,* the fixing of rates of charge, appointment of key staff, mapping and technical design, detailed investigation of materials position, location of stores, depots and determination of pay structures and levels.

The proposals were considered in detail by the Board over the course of seven meetings. Clarification was sought and given on several items. In a few cases the Board retained more direct control than was proposed but in general the proposals were agreed to with a small number of alterations. The agreed document thus became the organisational blueprint and was in no small way responsible for the speed and efficiency with which the Rural Scheme swung into action and achieved its objective.

RURAL ELECTRIFICATION OFFICE ORGANISATION

The first step was the setting up of the Rural Electrification Office (REO), the specific ESB management organisation charged with the carrying out of the scheme.[2] In charge was W. F. Roe who exercised broad control over the whole activity. He had been given by the Board a very high level of authority and discretion and delegated much of this downwards to the District and Area Engineers. His deputy and assistant was P. J. Dowling. Three Divisions concentrated on the main activities.

Materials Division was charged with securing all the materials for the Scheme in the quantities required and ensuring that these were delivered to the field crews when and where required. This involved the operation of a large transport fleet. It was also responsible for ensuring that the quality standards specified for the various items of material were achieved and at the best possible prices. Finally, the Division was charged with ensuring that the maximum amount of materials used was of Irish manufacture.

Recruited from outside the ESB organisation to head this Division was Patrick McDonald. He had been an engineer with Siemens Schuckert for two years on the Shannon Scheme and in 1929 had transferred to the ESB. In 1938 he had left to take up the position of manager of Siemens Schuckert Manufacturing Co. Ireland. Returning to the ESB his experience in the area of materials procurement for his previous company over the difficult war period was to be invaluable, particularly in the early post-war years when supplies of materials were still extremely scarce.

Technical Division was responsible for developing the most suitable electrical and mechanical design of the rural system, for the application of the most effective technology and construction methods and for the continuous assessment of standards of construction and productivity. In the early years of the scheme, very close liaison was required between Materials and Technical Divisions to cope with the recurrent shortages of many items by adapting and substituting.

Harold Montgomery was appointed as Head of Technical Division. He had worked in the Dublin Corporation Electricity Department and was one of the first engineers recruited by the ESB in 1927. After a period in Design Department he was transferred to District work. One of his particular responsibilities was the technical training of the many young engineers recruited for rural electrification. 'Monty', as he was affectionately called, pioneered many of the design and construction methods used on the scheme which contributed in a very large measure to holding down the capital costs during a period of severe price escalation.

Development Division was involved in almost every aspect of relations with consumers and with the public in general, with the setting up of the selection process, the economic return of areas, terms for supply and rates of charge. It also was responsible for investigation, development and promotion of the application of electricity in the home and on the farm.

Selected to head this Division in 1947 was Robert C. Cuffe. Starting as a graduate apprentice with Messrs Metropolitan Vickers, Manchester, he joined Design Department of the ESB in 1936, working in Lines and Main System Divisions. His work on the effect of lightning discharges on power lines won him his PhD. Having successfully established Development Division in REO, he moved to System Operation Department to head up its newly-formed Planning and Development Division. He was succeeded in REO by John Francis Bourke, known to all and sundry as 'J.F.', who in the early years of the scheme had been mainly engaged on the technical aspects, particularly the manual *Design of Rural Networks* which became the rural engineers' bible.

A prime task of the Development Division was awakening the interest of the rural community in electricity and motivating the householders not only to put it to use for lighting, but to raise living standards and to improve production on the farm. One of the earliest appointments made, therefore, was that of Publicity Officer for the Scheme. Patrick J. Ennis of Consumers' Department was appointed to the position in September 1946, charged with the development of the publicity and promotional aspects on a nationwide basis. Travelling to all corners of the country, he rapidly developed an effective set of relationships with the numerous

Completely at ease while working aloft, thanks to well designed and reliable climbing equipment.

voluntary and statutory bodies concerned with rural development. He was assisted by a very professional corps of demonstrators and lecturers who operated throughout the country during the course of the scheme, attending agricultural shows, area demonstrations and local functions.

Rural Accounts were integrated with the general Accounts organisation of the ESB, under the supervision of Divisional Accountant 'Neil' O'Donoghue who, while seconded whole-time to the Rural Electrification Office, still reported to the Chief Accountant. The particular and often unique circumstances of rural electrification frequently required special accounting consideration and O'Donoghue ensured that harmonisation was maintained with the established system while still allowing the new organisation to develop in a flexible and dynamic manner. He also became responsible for the chartering of shipping to transport poles from the Baltic to the Irish pole depots and to deliver materials by sea to rural areas around the west and north-west coasts.

The REO, though it was a separate and practically autonomous entity, was not completely isolated from other ESB Departments. If experience or expertise pertinent to the problems of the rural scheme were available in other Departments, they were freely sought and freely given. In this way, the total resources of the ESB organisation were marshalled for the benefit of the scheme.

THE DISTRICT ORGANISATION

It was at District level that the REO came face to face with the consumer. Staff under the control of the District Engineer addressed meetings, canvassed and organised, and submitted canvass results from areas for the selection process. When an area was selected it was construction staff accountable to the District Engineer who entered the lands to erect the networks and connect up the houses to the system.

At District Headquarters a Rural Organisation Engineer (ROE) was appointed as an assistant to the District Engineer solely for rural electrification. He supervised the construction crews, usually three to five, each of which was headed by a Rural Area Engineer (RAE). In addition, the ROE had a small staff engaged in the preliminary canvassing, organising of areas, and the preparation and submission of canvassing results for the selection. He ensured that the area crews functioned efficiently and that standards of productivity and workmanship were maintained in rural electrification work throughout the District.

THE AREA ORGANISATION

The area crew was the basic unit in the carrying out of the scheme. Areas, usually based on the parish unit, were of 25–30 square miles. As they came up in the selection process the area crew moved in, completed the work in four to six months and moved on to the next in line. At peak, forty such crews were operating simultaneously throughout the ten rural Districts.

The rural area staff was made up of a more or less permanent nucleus of various trades and skills which moved from area to area. Heading the crew was the Rural Area Engineer who was assisted by a Rural Area Clerk, a Rural Area Organiser and a Rural Area Supervisor with a number of linesmen and other skilled workers. In addition, from forty to sixty general workmen, in some cases up to one hundred, were recruited locally for the duration of the work in each area.

THE AREA OFFICE

The Area Office was the headquarters for all activity in the area itself. Initially the tendency was to select an office location in the geographical centre of the area near a crossroads and erect a prefabricated wooden building. Later on, however, experience showed that the advantages of a central location were outweighed by the better communications available if the office was sited near a post office or in a village where a telephone was available. The availability of a telephone was of great importance, as the recurring shortages of materials items required the RAE to be in constant touch with the supply situation so as to plan substitutions or re-schedule the construction programme quickly. In the immediate post-war situation subscriber telephones were scarce and telephone calls from rural post offices frequently took hours to complete. Much time was wasted if the telephone was any appreciable distance from the Area Office as this meant that no other work could be attended to while the connection was being awaited. In an effort to cope with the problem a number of radio transmitter/receivers was purchased from war-surplus stores. By modern standards these were very primitive in range and clarity, but they did give a limited communication with the District Office, which in turn could pass on or receive messages from Head Office. However, it was usually possible, with the help of the local committee, to secure offices and stores in a village where someone had a phone. The radio transmitters were phased out and a loud shout down the village street from the custodian of the local telephone brought the RAE – or more often the Area Clerk – running when the call eventually came through.

THE RURAL AREA ENGINEER

Recruiting to the engineering staff of the ESB had been virtually discontinued during the war years and, when the rural scheme was launched, very few mature or experienced engineers were available among ESB staff for transfer to rural work. The few who were available were of necessity allocated to the planning and organising work of the Rural Electrification Office. None were available for work in the field. In the post-war period not many experienced engineers were available from outside. It became obvious to W. F. Roe that if he was to meet the required timetable he would have to recruit and train young engineers, most of them fresh from graduation, to head the field crews.

This policy, born of necessity, turned out to be one of the strengths of the Scheme. What was required in an effective Rural Area Engineer was not so much maturity and experience as resourcefulness, common sense, energy and enthusiasm. These qualities were available in plenty in the young graduates of the fifties and sixties, many of whom went straight from college into rural electrification work. The reaction of an elderly parish priest in the west to this situation was typical of many. On the arrival of the construction crew, a member of the local committee introduced the engineer who was to be in charge of the whole operation. Obviously expecting a more mature man, the old P.P. blinked, sniffed and exclaimed 'Bless my soul! Has the boy been confirmed yet?'[3]

In the early years 'backsliders' were a serious problem. This term was applied to householders who had signed application forms but who when the crew arrived in the area had changed their minds. 'Backsliding' generally resulted in a loss of fixed charge revenue without a corresponding reduction in capital cost, so it was up to the RAE to try to recover the position. This involved the exercise of his persuasive powers – a subject unlikely to have been covered in his technological education. Nevertheless, young men[4] who would have perhaps been too shy to partake in a class debate in college quickly found themselves standing on public platforms arguing the case for electrification and countering objections and criticism from the floor.

While the responsibilities of the job were divergent and, for a young person fresh from college, formidable, there were a number of important factors in his favour. There was a strong element of crusade in the job – the improvement in the lot of the rural dweller – which appealed to youth and which was a very highly motivating factor. A high level of authority was given. The objectives and constrictions, financial and others, within which he had to operate were clearly set out. Apart from occasional visits from his superiors in the District or Head Office, it was left up to him to achieve the required results, but it was made clear from the outset that a high standard of performance was required. A poor performer could expect short shrift from W. F. Roe. The great majority of the young engineers thrived on this combination of high motivation and the high degree of responsibility and authority given to them. They tackled the job with enthusiasm and, with their equally highly motivated crews, achieved a rate of progress in rural electrification unparalleled in any country in western Europe, at a cost which justified the trust placed in them. That they and their crews also earned the respect and confidence of the community among which they worked is demonstrated by the fact that although over one million poles were erected, mostly on private land, the number of disputes and wayleave objections which were not settled on the spot but which rose to boardroom level could be counted on the fingers of two hands.

THE RURAL AREA ORGANISER

Above all, rural electrification had to do with people, the people of rural Ireland for whose benefit it had been conceived, to whose homes and farms it was being brought, over whose land 75,000 miles of line would stretch and on whose land over one million poles would be erected. In one way or another the scheme would have an impact on every rural housekeeper, householder and landowner in the country. It was therefore of the greatest importance that in each area under construction there should be an ESB officer whose responsibility lay in the areas of relationships between the Board and the people it was serving. As well as persuading potential consumers of the benefits of electricity this officer would have to measure houses and assess the fixed charges, get application forms signed, serve wayleaves and deal with objections in the first instance, organise demonstrations of electrical equipment, advise on selection and installation of electrical appliances and generally act as liaison officer between the Board and the consumer.

The recruitment of this new category of officer fortunately coincided with the large scale demobilisation of the country's war-time army. As a result very many well-educated and adaptable young men of both commissioned and non-commissioned officer rank applied and were taken on as well as young people of equal quality from other walks of life. A very short period of training sufficed to equip the new Area Organisers for their job as the advance guard of the scheme, helping to form local committees for the preliminary canvass and carrying out the subsequent official canvass on which the order of selection of the area was based.

When construction started the task of the Area Organiser was to check the final acceptance position and try to persuade householders having second thoughts not to backslide at this stage. The officer also had to serve the wayleave notices on the landowners over whose land the lines would run. These tasks were usually combined with that of ensuring that houses were wired internally when supply was connected so that it could be utilised immediately. With this constant contact with the householders the AO became the best known ESB figure in the area.

Simultaneously with these duties the Area Organiser was constantly promoting the use of electricity to improve the standard of living and to improve productivity on the farm, mostly using simple inexpensive methods in the early stages. At a later stage, the AOs visited consumers with sales vans fully equipped to demonstrate larger items of equipment such as pumps, grain grinders, cookers and refrigerators. The experience gained by the Area Organisers in dealing with people in different situations stood them in good stead in later years when the scheme was completed. Many of them became permanent salespeople and others developed into specialist advisers in various applications of electricity.

THE AREA CLERK

Unlike most members of the area crew who were recruited specially for the scheme, the Area Clerk was a member of the permanent accounting staff of the District, seconded to rural electrification for a period. This ensured that the accounting and stores control procedures in the area conformed to the standard procedures in force throughout the ESB organisation. The area pay-roll and stores control were the special responsibilities of the Area Clerk. As the RAE and AO were out 'on the line' for most of the day, the Area Clerk was the only officer normally available in the Area Office. This being so he also found himself involved in a multitude of enquiries from callers in the course of the day. It was usually he who had to ring Head Office on the local telephone to query the latest material shortage and discuss possible alternatives or go sprinting down the village street in answer to a shout from the custodian of the phone when they called back. When aggrieved consumers or landowners called to the office, it was usually the Area Clerk who had to bear the first impact of their wrath. Being involved so intimately with the custody of materials, their shortages and substitutions, the Area Clerk developed a detailed knowledge of all hardware used in line construction, the various pole sizes available, where they could be used etc. and so was an invaluable source of reference.

Far from being a mere upholder of approved accounting procedures, the Area Clerk was involved in almost every aspect of area work and developed a very good relationship with the rest of the crew, agonising with the AO over backsliders, sympathising with the RAE when materials ran short, and rejoicing with them all when the first consumer was switched on. Many of these people who subsequently went on to occupy senior accounting and administrative positions in the ESB organisation still look back on their spell of rural duty as having given them a special understanding of country life and living.

THE AREA SUPERVISOR

In direct control of all crew members actually engaged on construction work was the Area Supervisor. In the early phases he was sometimes an electrician or more often a senior linesman seconded from the District construction and maintenance staff. Whatever his background the Area Supervisor was selected for proven leadership qualities. It was he who had to weld a usually inexperienced group of people into an efficient working unit, employing a blend of carrot and stick.

Usually in the early stages of the Scheme the Area Supervisor had the help of one or two experienced linesmen and a few semi-skilled men who in turn trained some of the more promising of the locally-recruited general workmen into the more skilled aspects of line work. These in turn took charge of smaller units of unskilled men who dug holes, erected poles, strung lines, erected transformers and serviced houses. The locally recruited trainees were invited to move with the permanent crew to the next area and so developed into skilled linesmen. Over the period of the scheme there was a constant renewal upwards of the various levels of skills as young recruits moved from unskilled to semi-skilled, to skilled, to Linesman, to Chargehand Linesman and in many cases to Area Supervisor or to senior Line Supervisor positions on the District staff. One of the attributes of a good Area Supervisor was the ability to detect qualities in young recruits which would fit them for the promotional ladder.

The four officers described, Area Engineer, Organiser, Clerk and Supervisor, together with a small number of skilled and semi-skilled workers – about ten to fifteen in all – formed the nucleus of the rural area crew and travelled from area to area. In each area, local men were recruited for the less skilled work in excavation, pole erection etc., to form a total crew usually of forty to eighty or sometimes even up to 100. A crew of fifty could complete an area of four hundred consumers in about five to six months. At the peak of construction, which was reached in 1956, the total number of crews engaged in the work had reached forty. In that year over 10,000 kilometres (6,000 miles) of line were constructed involving the erection of 112,000 poles. Ninety-nine areas were completed and over 34,000 new rural consumers connected.

W. F. Roe, left, first Engineer-in-Charge of rural electrification and P. J. Dowling, his successor.

CHAPTER SIX

REO News

IN TACKLING what was then one of the largest development projects ever envisaged in the country, W. F. Roe was keenly aware of the importance of good communications with staff. If a high standard of performance was to be achieved, the staff needed not alone to be well briefed and motivated at the start, but to be constantly refreshed with information on the progress of the scheme, advised of developments in all aspects of the work, sustained when difficulties arose and motivated to give of their best at all times.

As the peak of the scheme would find forty separate working units each of from fifty to one hundred people of various disciplines and skills scattered throughout the 26,000 square miles of the State, some in very remote localities, day-to-day personal contact was out of the question. In an intensely busy headquarters there was little time or opportunity for developing a communications organ. W. F. Roe, recognising the importance of such communication from an early stage, decided that a start, however imperfect, would have to be made. One day in December 1947 in a typical gesture he called in a typist and, in his own words, 'dictated the first issue of *REO News* from cover to cover', three foolscap pages which were then issued in stencilled form. The first paragraph of this first issue reads:

> In order to keep the rural staff informed of the progress of the Rural Electrification Scheme, it is intended to issue *REO News* monthly. The date of issue will be towards the end of each month when progress has been reported to REO from each area and District. General information regarding the various aspects of the scheme such as suggested construction methods, general position of the supply of materials, and applications of electricity to agriculture will be given and perhaps a social column so that you will have some information about your friends in various parts of the country. Contributions from members of the Area or District staffs will be welcomed.

The issue then went on to give an overall review of the scheme. Areas had now been selected in twenty-three counties and it was hoped to have selections in the remaining three – Cavan, Monaghan and Donegal – before the end of the month. With regard to materials, pole supplies were reasonably secure with the purchase of 114,000 poles in Finland, of which 95,000 had been delivered before the ice closed the Baltic ports. These would supplement the much smaller quantities of Irish poles that were available. The situation regarding other key items of material was then gone into in some detail.

From small beginnings *REO News* blossomed into a sizeable monthly of about twenty pages whose arrival was awaited with eagerness by the field and district staff. After the first issue, W. F. Roe succeeded in delegating responsibility for its production while retaining his prerogative as Editor in Chief. Effectively the editorial responsibility was placed on the shoulders of John Francis Bourke, assistant Head of Technical Division. In 1950 he succeeded R. C. Cuffe as Head of Development Division. There was a polite fiction that the identify of the editor was an official secret, but even the rawest recruit was not left long in doubt about the authorship of the pungent editorials and the pithy snippets that laced the pages of the *News,* making it sometimes controversial, sometimes pontificial, but never dull. J. F. – he was never known except by the initials – remained editor and producer of the magazine for almost its full fourteen years life.

The magazine retained its foolscap format for the first twelve issues and then changed to the handier quarto size. Costs of production were kept to the minimum and for twelve years it was produced in stencilled form using the normal typing and duplicating facilities of REO. An exception was the December issue, which was regarded as the Christmas number. This was a fully printed product and was more light-hearted in tone. 1953 saw an innovation in a printed cover page of good quality glossy paper which enabled photographs of rural electrical appliances to be reproduced and gave the publication a somewhat more professional look. Finally, for the last two years of its life, *REO News* appeared in an attractive fully printed form which made it possible to include a wider range of material.

In the course of its life – from December 1947 to November 1961 – *REO News* was to play a most important management role in informing, educating and motivating the widely dispersed staff, in countering their sense of isolation and in building up a team spirit. It provided a vehicle for the often spirited exchange of views, for criticism of performance of management and field worker alike. It acted both as a suggestions box and as a safety valve as its columns were open to all rural staff to make

'The old order changeth...'

This is the last issue of " R.E.O. NEWS ".

In December, 1947, the first number of the NEWS made its modest appearance " in order to keep the Rural staff informed of the progress of the Rural Electrification Scheme ". For fourteen years it has fulfilled that promise.

With the end of the area development programme next year over 280,000 rural consumers will be connected to our networks at a cost of over £30,000,000. This was no small achievement. It was not always easy, but the co-operation and enthusiasm of the staff carrying out the Scheme in the areas or planning and helping it in the District and Head Offices overcame all obstacles. Now the job is almost done and " R.E.O. NEWS " has outlived its usefulness.

In the second issue of " R.E.O. NEWS " (January, 1948) the following note appeared :—

" Quidnunc " discussing the memorable events of 1947 closes his column in the " Irish Times " as follows :—

> *" But how many of these things will be remembered in, say 2047 ? I dare swear that if any event is recorded in the history books (taught through the medium of Russian) it will be none of those I have mentioned ; rather, it will be one which has passed almost unnoticed, amid the turmoil of the year.*
>
> *Somebody—I cannot remember who—switched on the lights in some village—I cannot remember where—and rural electrification took her bow. And if that does not mean more to the country than all the rest of the year's events put together I shall be very surprised indeed."*

No comment is required. Though the Scheme is not quite finished " R.E.O. NEWS " bows out with pride.

T. J. Dowling

suggestions and criticisms or simply to sound off and get rid of a head of steam. In this way it provided a very effective feedback to management.

Through its leading articles in particular the magazine focussed the attention of the staff on the broad aspects of the work. It praised or criticised performance and exhorted all to greater efforts. It highlighted areas where it felt improvement was required – construction costs and standards, work output, vehicle mileage, accidents, appliance sales, backsliders etc. In the area of work output per crew it was frankly stakhanovite in its approach, setting up methods of comparing the performance of the different crews and so hoping to inculcate a sense of rivalry. The first step was to establish a way of measuring work and of expressing amount of work done, however different in character, in terms of comparable 'work units'. The monthly reports from the areas showed the number of work units done and thus, month by month, the performance of the crews both in output and cost could be compared.

A good average-sized crew working well could achieve over one thousand work units in a four-week month and so the idea of the M areas (M = 1,000) was developed. Each month the M areas were listed together with the names of the engineer and supervisor concerned together with suitable words of praise. While non-M areas were not mentioned, they were of course conspicuously absent and motivated, it was hoped, to earn the accolade at a future date. If a crew either larger or smaller than average achieved M status the praise was muted or heightened accordingly.

An item that proved popular was the Top Ten where the cumulative performance of each crew (in terms of work units) from the beginning of each year was logged. Each issue of *REO News* gave the placings of the ten crews which currently had the highest cumulative outputs. The crews and their placings changed from month to month for various reasons not all under the control of the RAE. The prevalence of rock, for example, could increase costs and reduce output. Similarly, completing one area and transferring to the next could affect work output for the month in which the change occurred. The placings at the end of the twelve months were, however, generally accepted as a fair measure of each crew's efficiency.

While this may appear rather naive in these days of more sophisticated management techniques and, indeed, more sophisticated worker-response, it was effective in keeping the importance of high productivity constantly to the fore. The Top Ten was invariably one of the first features to which engineers and supervisors turned on receipt of their copy of *REO News*.

Each of the three Headquarters Divisions – Materials, Technical and

Development – was given space in the journal to keep staff up to date in its particular area.

MATERIALS DIVISION

In the first few years of the scheme in particular, materials for construction were difficult to obtain and deliveries were sporadic in the extreme. This made the achievement of high productivity in the field difficult and sometimes impossible. If, however, the RAE was kept in constant touch with the materials position – which materials were likely to go short and when, or whether deliveries of some materials in short supply could be expected and when – he could do much to counter the effect of the shortages by the suitable re-deployment of his manpower on the ground. The Materials Division Notes, which appeared in each issue of *REO News* and which gave the most up-to-date position on each item of stock were therefore of great value in allowing the most efficient planning of the crews' time and thereby keeping up the momentum of the activity.

TECHNICAL DIVISION

The notes provided a continuous updating of the basic design manual, *Design of Rural Networks* of which every RAE had a copy. New materials coming on the market, new technology, practical construction problems met in the field and their solutions, improvisation and substitution for scarce materials were some of the subjects treated. A random sample from just three of the over 160 sets of Technical Division Notes gives the following:

Balancing of single phase loads, tree cutting, undesirable servicing, faulty S.C.A. conductor, service poles.

Binding-in S.C.A. conductor, rural system improvements, bolted and compression connections.

Records of P.O. crossings, new eaves fitting, H.T. construction on L.T. lines, damage to roofs, earthing in rocky country.

DEVELOPMENT DIVISION

Materials and Technical Divisions had to do with the physical construction of the network. Development Division focussed on the consumers themselves. It monitored the economics of whole areas and individual extensions but the greatest part of its efforts went to ensuring that once electricity was made available it was utilised to the greatest possible extent. Demonstrations, exhibitions, participation in shows, educational activities, liaison with rural voluntary organisations, development of applications of electricity to the rural home, to providing water on tap, to agriculture and horticulture, promotion and sale of appliances – all these activities and more came under the umbrella of Development Division. Consequently its notes covered a wide field including such items as, How to Measure a House on Your Own, Reports on Shows, Sales campaigns, Conversion of Milk Separators to Electric Drive, Test Reports on Appliances, Diagnosis of Pump Troubles and What the Salesman Should Know about Welders. If these notes could be said to have one recurrent theme it was the necessity for selling as much suitable electrical equipment as possible to the newly connected consumers so as to ensure that from the start they would reap the maximum benefits from electrification.

FEEDBACK

A very important role of *REO News* was to provide a forum in which the men in the field could exchange their experiences of dealing with problems, criticise and make suggestions for changes in policy and generally provide feedback to management about how the scheme was progressing. This was of particular value in monitoring public reaction to the scheme as the people in the field were in daily touch with a broad cross-section of the community.

PROGRESS

The progress of the scheme was faithfully recorded in the pages of *REO News*. As work commenced in each area, the number of poles erected, length of line strung and consumers connected were recorded month by month until the note 'completed' was entered and the crew moved on to their next location. The grand total to date was also given so that all readers were kept up to date on the overall picture.

SOCIAL

Finally and most importantly for staff members, the comings and goings of their colleagues were chronicled. Newcomers to the staff were welcomed by name, transfers from one area to another or to work areas outside 'rural' were meticulously recorded. Engagements and marriages were given full coverage, often in intimate and exquisite detail. When in the course of time the marriages bore fruit, the newcomers were welcomed into the world. Thus *REO News* contributed to building an *esprit de corps* which persisted long after the disbandment of the REO organisation.

In November 1961 the last issue of *REO News* appeared. It had been founded 'to keep the Rural staff informed of the progress of the Rural Electrification Scheme'. Now the pioneering work of initiating the rural community into the benefits of electricity had been successfully completed and rural dwellers were approaching the sophistication of their urban cousins in things electrical. The task of *REO News* was done and it marched out with banners flying to make way for a new magazine, *Prospect,* which would encompass a wider marketing spectrum under the editorship of Michael V. O'Connor.

CHAPTER 7

Material Matters

THE PERIOD IMMEDIATELY AFTER THE WAR was not the most propitious time for obtaining the wide range and large quantities of materials required for a nation-wide rural electrification scheme. There had been no opportunity to accumulate stocks during the war and in the initial years, keeping the construction crews supplied was a major problem. While a small number of items – poles, conductor and transformers – accounted for the bulk of materials (about 80% of the cost), there was a very large range of other items, the absence of any one of which would prevent the completion of the networks. Insulators with their pins and suspension clamps, crossarms, stayrods and staywire, fuses and fuse isolators, connectors and even nuts and bolts were just as vital. The position was aggravated by the recurring failure of most suppliers (they had their own difficulties) to deliver even reasonably close to the promised dates. Some of the shortages would have had a far more serious effect if the Rural Area Engineers had not shown their ability to adapt quickly to the ever-changing materials situation by rearranging the work of their crews.

After the first couple of years the supply situation improved, but so also did the demand as the number of crews was stepped up. It can be said that at any point in the course of the scheme there were always some components in short supply and much time and ingenuity were devoted to developing substitutes or in rationing out available supplies according to greatest needs. Very often unofficial bartering sessions were held between neighbouring Rural Areas as items not immediately needed were traded for those without which work would be held up.

POLES

The item that accounted for the greatest cost was poles. It was estimated that over one million poles would be required and that when

Linesman, aloft, prepares for the stringing.

construction reached its peak, over 100,000 poles per annum would be erected. (In fact, in the year 1955/56 112,616 were erected.) A thorough search for suitable wooden poles was undertaken in Ireland both in State forests and in privately owned plantations. Some suitable supplies of larch and Scots pine were located, generally in small parcels in woods scattered throughout the country from Killeshandra in County Cavan, to Tallow in County Waterford. By the end of 1945 a total of 6,000 poles of Irish origin had been ordered. This, however, was only a small fraction of requirements and it became obvious that if the scheme was to be constructed using wooden poles, the bulk of these would have to come from abroad.

A possible alternative to wood as a material for poles was reinforced concrete as used extensively in some European countries. The government was most anxious that this alternative should be examined as it feared that sufficient wooden poles would not become available, even from abroad, to commence the scheme. In the case of reinforced concrete poles, on the other hand, apart from the reinforcing steel, all materials required could be supplied from native sources. A detailed costing exercise was carried out, which indicated that on the basis of post-war prices, the use of concrete poles would add about £7 million to the cost, i.e. £28 million as against £21 million.[1] A small number of concrete poles was manufactured by Moracrete Ltd., Crumlin, Co. Dublin, to gain practical experience of manufacturing and erection costs. These were erected at Fairyhouse in north County Dublin and in Ballytore rural area in County Kildare, where they continue to give good service. However, their weight made them difficult and expensive to handle and the price differential remained and still remains in favour of wooden poles. Nowadays, however, a high proportion of the wooden poles required is supplied from Irish forests.

Source of Supply □ The search for wooden poles was extended to Scandanavia. Timber was available in Sweden and Norway, but in the post-war period most of it was earmarked for home use and the governments concerned were keeping a tight rein on exports. Sweden in particular was curtailing exports to Great Britain because of failure to obtain reciprocal imports, and so in that country sterling was not a very popular currency at the time.

The Swedish allocation in 1946 for all pole exports to Ireland, including P 7 T requirements, was for between 5,000 and 10,000 poles at a very high price. Norway had fairly large quantities on offer, but again the price was unacceptably high. Finland offered the most favourable prospect of obtaining sufficient poles at a reasonable price. Offers were received through agents and by May 1946 16,700 Finnish poles had been ordered to supplement the 6,000 obtained at home. This meant that supplies for about sixteen areas were now secured.

In view of the huge quantities involved it was decided to make direct contact with the principal Finnish pole suppliers and ascertain their ability to meet large orders. In August 1946 Neil O'Donoghue travelled to Finland. He found that there was ample timber suitable for poles standing in the Finnish forests. The Soviet Union had first call on all Finnish products as war reparations, but at this time it appeared that it had sufficient supplies of poles and was concentrating on manufactured goods, ships, machinery and prefabricated housing. For this reason, Finland's ability to supply poles to Ireland was expected to be good. Moreover, Finland was most anxious to sell for sterling, which of course was the Irish trading currency of the time.

O'Donoghue made contact with the Finnish Pole Exporters Association, a body set up to deal with the export of poles to countries with which Finland had concluded agreements during the war years and which was intended to function during the reparations period. Four members of the association, who between them handled 70% to 75% of all pole exports, arranged to deal with REO as a group. They agreed that one of their number, Mr Onni J. Salovaara, would, as well as dealing on his own behalf, act as agent for the other three with regard to all shipping matters and correspondence.

After some days of intensive negotiation, agreement was reached for the supply of 90,000 poles at a satisfactory price (averaging about £2 per pole). W. F. Roe received a telegram from O'Donoghue advising him of the final asking price and requesting permission to close the deal. He decided to buy them but had a problem in conveying his agreement to O'Donoghue without compromising the latter's chances of improving further on the price in the final sealing of the bargain. The difficulty was that the only telegraphic address for O'Donoghue was the Salovaara Office in Helsinki, so that any indications of satisfaction with the price would immediately become known to the Finns. Not for the first or last time the native language was called upon to help out. The telegram from Dublin read: 'Praghas maith go leor, Roe' ('Price satisfactory, Roe'). O'Donoghue solemnly informed the group that they would have to improve on the price to get the contract, which after some consultation they did. O'Donoghue immediately telegraphed his boss: 'Have done deal. Get Board approval in ainm Dé' ('. . . in God's name'). The first major contract for the scheme had been successfully negotiated. Subsequently the Finnish suppliers found it possible to offer a further 24,000 poles as an extension of this contract making the total quantity purchased for delivery in the 1947 shipping season to 114,000.

While other Finnish suppliers subsequently came on the scene , the firm of Onni J. Salovaara and its associated companies remained major suppliers of poles for rural electrification and a relationship of the highest mutual respect was established which survived throughout the thirty years of the Scheme. Many years later, Mr Salovaara got around to asking about that first telegram from Dublin – the Finns had tried all the standard international telegraphic codes without success – and was highly amused to find that he had been stymied by the Irish language.

At all times preference was given to the purchase of Irish poles but these were never available in anything approaching the quantities needed; so up to one million poles were imported, almost entirely from Finland. They were purchased at very keen prices, firstly because of the large quantities required, and secondly because the REO conducted business directly with principals and not through middlemen. A further attraction to suppliers was the small sizes and assortment of poles that were acceptable: by international standards the pole sizes required by REO for rural work were generally on the small side; furthermore, REO allowed a good latitude in regard to the numbers of the various sizes acceptable to it. The result was that the pole producers did not have to do so much 'picking' in their felling operations and found it easier to handle the smaller sizes involved. This was reflected in the prices quoted.

While there was a strong element of competition among suppliers for REO business, the Finnish government made sure, through a system of export licences, that poles were not sold below what it considered a fair price. On some occasions, in an effort to get business, small suppliers quoted very low prices to find that they were unable to get export licences. On other occasions, it was found necessary to split orders between a number of suppliers to ensure that licences covering the required total were granted.

Inspection of poles when felled was a major task. REO insisted that every individual pole should be inspected and passed before shipping. Initially this was done by REO's own forester, Dermot Mangan, who spent many long winter weeks in the snowbound Finnish forests travelling by train, bus, taxi and sleigh, inspecting poles which were sometimes buried in snow and at other times 'had as much as an inch of melted snow on them in the form of solid ice which made the measuring and calipering no joke'. Eventually, as the quantities to be inspected grew rapidly, a native Finnish forester, Viljo Rantala, was given training in Ireland on precise REO requirements and returned as REO pole inspector to Finland. This arrangement proved very satisfactory: there were even some complaints from his fellow countrymen that he was too rigid. Paddy McDonald, who was responsible for all material purchases, averred that 'as a result, we practically never got a bad pole delivered out of a total of one million'.

The first cargo of Finnish poles left the port of Hamina for Dublin at the beginning of July 1946 on the decks of the S.S. *Ashbel Hubbard*. This was followed immediately by 2,000 poles as deck cargo on the *Irish Larch* and in the same month the S.S. *Wicklow Head* was reported as on her way to Hamina to lift 13,000 poles for delivery to Cork. From then on and for the next thirty years there was virtually never a time during the 'open water' (ice-free period from May to December) in which pole-carrying ships for rural electrification were not either loading in Finnish ports, making the long voyage from the Baltic to Irish waters or unloading at the three REO port depots of Dublin, Cork or Limerick.

Creosoting and Storage □ The purchase of sufficient poles in bulk was of course only the first link in the chain that terminated in serried ranks of rural electrification poles marching across the Irish countryside. Bulk storage and treatment depots had to be set up in Ireland and these were established in Dublin, Cork and Limerick. The locations were selected for their port reception facilities and for their strategic location with regard to the distribution of the poles to the rural areas. In Dublin a large area was rented from the Port and Docks Board at East Wall Road convenient to the creosoting plant of Messrs T & C Martin. In Cork the reception and storage area was on the river adjacent to the Ford Plant and again convenient to an existing creosoting plant operated by Messrs Eustace & Co. Creosoting as a method of treating raw poles for preservation was used practically exclusively. From time to time other preservation methods were tried, such as various types of salts, but none was adopted for general use. By 1947 it was necessary to open a third depot at Limerick port and a new creosoting plant was put into operation by Messrs Eustace through the Limerick firm of Spaights. Through these three depots and treatment plants passed all the poles for the scheme and apart from some problems in the immediate post-war years when supplies of creosoting oil dried up periodically, there was never any serious delay in the supplies of creosoted poles to the areas.

Delivery of creosoted poles to most areas was affected by a fleet of tractors and specially constructed pole trailers, but in the case of areas in County Donegal and parts of the west coast small coasters were chartered to transport the complete package of materials required.

CONDUCTOR – 'THE WIRES'

The original design for the rural networds was mainly for copper conductor and by October 1945 British firms were once again open to orders for this. It quickly became evident, however, that the rapidly rising price of copper would seriously escalate construction costs and alternatives were urgently examined. These included aluminium, aluminium alloy, steel-cored aluminium and cadmium copper. Early in 1947 the very important decision was made to use steel-cored aluminium (SCA) almost universally, with all-aluminium conductor (i.e. without the steel core) for some of the low voltage networks where, because of short spans, the mechanical tensions were low. In the case of small, lightly loaded, high-voltage spur lines, it was intended to use galvanised steel stranded conductor, but the acute post-war steel shortage made this very difficult to obtain. Light gauge SCA was substituted with satisfactory results.

The first orders for the aluminium conductor were placed with the Aluminium Union of Canada and deliveries commenced in 1948. Because of the acute dollar shortage there was some difficulty in getting clearance to place the orders with the Canadian company, and in his regular monthly report to the Board in January 1948 the Engineer-in-Charge noted that 'the first consignment of SCA from Canada is at sea. Permission to place the second order with Canadian suppliers amounting to $160,000 [Canadian] was only obtained from the Department of Finance at the latest moment to hold our position on the supplier's order book.'

The Aberdare Electric Co. Ltd. □ Alternative sources of supply were located in Britain, Belgium and Germany, but at this stage an Irish supplier had entered the scene, the Aberdare Electric Company, later to develop into the Irish industrial giant of Unidare Ltd. When, in 1944, the government's intentions regarding rural electrification were announced, the possibilities of providing some of the equipment by an Irish company were investigated by a number of Irish businessmen who founded a consortium for the purpose. The Aberdare Electric Company was launched in November 1947 with an issued capital of £210,000.[2]

The initial technical know-how regarding the manufacture of overhead electrical conductor and transformers was provided by bringing on to the new Board two members of the Board of Aberdare Cables Ltd. of South Wales – hence the name of the Company. By the mid-1950s, the activities of the Company had expanded into many fields outside the electrical industry. For this reason and to avoid confusion with the Welsh company and to emphasise its independence, the title of the Irish company was changed to Unidare Ltd.[3] It has long diversified from its initial concentration on electrical equipment, but it freely acknowledges that its springboard was the considerable initial market for its products provided by the Rural Electrification Scheme, particularly transformers and overhead conductors. Later came the manufacture of covered conductor and underground cables, storage heaters, water pumps and piping and a large range of accessories for line construction all developed in very close liaison with the ESB.

The original machinery for the manufacture of conductors was designed for copper so that when the price rises caused a switch to steel-cored aluminium conductor there were problems in converting the plant. A very big problem was the welding together of lengths of high tensile steel core for the SCA conductor. This was overcome and in August 1949 the company received an order for half the year's requirements of SCA for the scheme. In addition, the ESB gave the company a further three months to negotiate for further supplies of aluminium bars in order to be in a position to quote for the balance.

The devaluation of sterling in 1949 had a serious effect on the price of raw materials for the conductor manufacture. Copper increased in price from £107 per ton to £140. With regard to aluminium the monthly Rural Electrification Report for September 1949 notes: 'The British Aluminium Metal Price has increased from £93 per ton to £112. Fortunately, the Aberdare Company had purchased the aluminium for our last order before devaluation so that we are covered on conductor for about nine months ahead.'

TRANSFORMERS

After poles and conductor, the third important item of material was transformers. Owing to the distances involved and the dispersed nature of the dwellings the most economic method of distribution was at 10,000 volts (10kV) mostly single-phase. To reduce this voltage to the domestic working voltage of 220 volts required transformers. Unlike the urban situation where one large transformer – 400kVA or 640kVA [kVA—Kilovolt-Amperes) for example – would serve several hundred premises, the rural transformers were generally expected to serve groups of from one to ten houses with four or five houses being the most usual number. This necessitated a very large number (up to 100,000) transformers of small size. The most common sizes in the initial stages were 2½kVA, 3kVA, 5kVA, 15kVA and, for groups larger than normal, 33kVA.

American units were initially the cheapest available and considerable orders were placed with Westinghouse in the U.S. As with the conductor, however, the dollar shortage and the devaluation of sterling in 1949 made the American equipment expensive and difficult to obtain. Enquiries in Great Britain resulted in quotations of almost double the American price and there were some indications that a British 'ring' was in operation. The search was extended to Belgium, Sweden, Switzerland and Czechoslovakia without very much success.

By the end of 1946 orders for 1,500 transformers had been placed with American and British firms, and it was expected that about one half of these would be delivered by the end of the following year. The difficulties in obtaining the American transformers and the slow delivery and very high price of alternative British and Continental supplies had raised the question of manufacture in Ireland. One British manufacturer (Brush) had already made overtures. By mid-1947 the prices of American transformers had increased to almost the British level and the delivery position was still difficult.

In May 1948 an order for five hundred 5kVA transformers was placed with South Wales Switchgear, an associated company of Aberdare Cables Ltd. On 7 November 1947 the Aberdare Electric Co. Ltd. had been formed in Ireland to establish and develop a large-scale electrical equipment manufacturing industry with two members of the Board coming from the board of Aberdare Cables Ltd. By 1949 the company was in a position to quote for transformers and in November of that year the ESB placed their first order with it for three thousand 5kVA transformers, about twelve months requirements. Manufacture and deliveries commenced immediately and on 22 December the first Irish-made transformer was erected in the Man-O-War rural area in north County Dublin. It is well to record that in the development and testing of these transformers there was a very close liaison between the Company and the ESB thus ensuring high quality and close matching to ESB requirements. The amount of Irish manufacture (as distinct from assembly of already manufactured parts) was very high; tanks were rolled and welded from the sheet, shot-blasted and spray-painted; cores were assembled from stampings; coils were wound on the premises. In fact, the transformer lid casting and the bushings were about the only manufactured parts imported. The first batch of transformers was delivered in January 1950.

ACEC (Ireland) Ltd. □ In pre-war years the Belgian Company, Atelier de Constructions Electriques de Charleroi (ACEC), had done business with the ESB. When Paddy McDonald visited Belgium and Holland in October 1946 in search of materials to enable the scheme to commence, he found that ACEC was just recovering from the disruption of the war with most of its immediate output designated for urgent home requirements.

Nevertheless, Jean Kottgen, chief of the export department, expressed an acute interest in the long-term market for transformers and electric motors which the rural electrification scheme was opening up in Ireland. The upshot of the meeting was a more thorough investigation of the Irish market by the company which noted particularly the Government stipulation that as far as possible, Irish materials should be used on the scheme. This resulted in the establishment in 1951 of ACEC (Ireland) Ltd., a wholly owned subsidiary of ACEC, S.A., of Belgium, with a factory at Tycor in Waterford City, to manufacture transformers both for the rural electrification scheme and for general distribution requirements. Like the Aberdare Company, established four years earlier, ACEC had a company policy to manufacture to the maximum extent rather than just to assemble. In fact the only finished bought in components were bushings and tap changing mechanisms.

This policy ensured not alone in high level of local employment but laid the foundations for subsequent expansion into a much larger and wider range of products. Initially, concentration was on 3kVA single-phase transformers for the rural scheme and the first of these was erected in St John's rural area in County Kilkenny in July 1952. The range of transformers manufactured was extended gradually over the years to 38kV/10kV and then to the more recent 110kV/38kV giants of 31.5MVA (31,500kVA) capacity. Again as with Unidare, the very close liaison and co-operation established in the early days of the rural scheme has been maintained to the benefit of both organisations and of the Irish electrical industry in general.

For a period, the Waterford factory manufactured electric motors of up to 30hp in three-phase and up to 3hp in single phase. These latter were designed specially for rural conditions and many thousands were used to provide motive power for pumps, grain grinders and milking machines in rural areas. However, with the lifting of trade barriers, competition from large-volume foreign manufacturers resulted in the Irish motors being priced off the market and the main product apart from transformers now coming from the Waterford company is electrical flourescent fittings. These are produced in a modern plant in the factory complex and account for a substantial part of turnover. In 1982 the company employed three hundred people.

OTHER MATERIALS

Apart from the three major items – poles, conductor and transformers – there was a host of other items, the absence of any one of which could hold up work. These included ironwork of various kinds, struts, ties for headgear, insulator pins, suspension clamps, shackles, shackle straps, earth rods, stay rods, and a myriad of nuts and bolts of various sizes, connectors of various kinds, dead-end thimbles, insulators for H T (high tension) and L T (low tension) fuses, air break switches.

Immediately after the war, manufacturers, especially those in Europe, had to face a major task of rebuilding or re-adapting their plants for peace-time work and of locating supplies of raw materials, very often from totally new sources. Like so many other industries the electrical industry had to undergo a difficult period of re-adjustment which took several years. Even ten years after the war there were recurring shortages of various items of material, which sometimes caused serious hold-ups in construction work. A compensating feature was the large stock of war-surplus material released on to the market. While this had been produced for the war effort, it was possible, by intelligent adaptation, to use much of it for peace-time purposes. A good example on the rural scheme was the use of British Admiralty surplus minesweeping hawser as a staywire. Owing to its hardness the multi-strand hawser could not be spliced in the conventional manner. An alternative technique was evolved using wire rope clips which permitted the use of this immensely strong but difficult-to-handle material as a substitute for stay wire. About eight hundred miles of hawser were used in this manner before supplies of conventional stay wire were resumed. W. F. Roe recalled later that just as delivery had been completed, an enquiry came from the British Admiralty to ask if half the consignment could be returned. Obviously its adaptability for uses other than minesweeping had been discovered. Needless to say, he was not disposed to meet the request.

The procurement of sufficient materials to allow a start to be made necessitated numerous visits to manufacturers in Great Britain and on the Continent as well as to the many war-surplus depots which were springing up as the huge wartime installations were decommissioned. There were some initial difficulties for a country which had been neutral in the conflict. In Britain itself some degree of reserve and even hostility was encountered. Ireland's neutrality had been criticised and misrepresented by Winston Churchill in his widely disseminated victory speech to the nation. De Valera's reply had got little publicity in Britain and Paddy McDonald, the man responsible for procuring the materials, recalled that in his early meetings with suppliers much time was taken

'Made in Waterford' — The first batch of 250 3kVA rural electrification transformers manufactured by ACEC awaiting collection, June 1952.

up with remedying this omission. Subsequently, excellent relations were developed with many key British suppliers who, being first and foremost keen businessmen, looking to the future rather than the past, recognised the large potential export market provided by a scheme of such magnitude and made special efforts to meet REO's requirements.

On the Continent the search was no less intense. On one trip alone, in late 1946, McDonald visited no less than thirty-six Dutch and Belgian firms seeking supplies of copper and steel conductor, galvanised steel headgear and accessories, insulators, insulated cable, transformers and switchgear. He had limited success, one firm hesitating to quote when asked if it would accept sterling in payment. His assessment was that the supply position would not ease out until 1948. However, through persistence and adaptation, sufficient materials were procured to allow the scheme to get under way as planned.

Gradually, as more Irish firms started manufacturing the proportion of materials obtained from Irish suppliers increased. Indeed in some cases the spread of rural electrification itself enabled Irish-made products to be offered. A shortage of steel for crossarms – used by the tens of thousands – necessitated the substitution of native oak crossarms, but these were obtained only with difficulty until the development of the Dundrum rural area in County Tipperary, early in 1951. The Department of Lands installed a sawmill powered from the new rural electricity network in the State forest which had good stands of suitable oak. Ample supplies of oak crossarms of good quality were soon on their way.

It was the conscious policy of the REO to purchase as much as possible of its material from Irish suppliers and to encourage in every way possible the setting up of new Irish industries to meet its needs. By 1957 the Board was able to report that for the year 1956/57, out of £2,090,000 spent on rural electrification materials, £1,442,000, or 69% had been purchased from Irish suppliers. By 1961 this had risen to 82% and by 1965 to 84%.

CHAPTER EIGHT

Shipping the Poles

THE BALTIC TRADE

THE SHIPPING OF ONE MILLION POLES was in itself a huge operation. At the outset, the Rural Electrification Office staff knew very little about the niceties of ship chartering. The necessity to have huge quantities of poles shipped promptly and cheaply provided the incentive to learn quickly. Shipping from the Baltic was a particular problem, as from early December to about the middle of May the ports were icebound. Shipping therefore had to be compressed to a seven-month season, starting with 'first open water', which generally occurred during the first or second week in May.

Poles are awkward to stow and not every ship is suitably constructed to carry them, particularly under decks. In earlier days, when the amounts were comparatively small, poles for the ESB were generally shipped as deck cargo after other cargo, usually sawn timber, had been stowed in the holds. Another difficulty was that poles bulk much more than an equal weight of sawn timber. The normal shipping unit for timber in the Baltic trade is the 'standard', (165 cubic feet of sawn timber). In the case of poles, however, one can only get about two-thirds of this amount of actual timber into the same volume, so that only ten average-sized poles, totalling about 110 cubic feet, go to make up a standard. As time went on and the ESB's requirements became known in Baltic shipping circles, ships specially suited to the pole-carrying business were offered. As these could carry larger quantities both in the holds and on deck they could be chartered at more reasonable rates.

In the initial stages the easiest terms to obtain were 'liner' terms, whereby a ship on a scheduled run between ports would carry a quantity of poles – usually on deck – as part of a general assortment of cargo. For small quantities this was quite suitable and the most economic method. The first shipment of poles from Finland for the Rural Electrification

Scheme was carried on this basis by the S.S. *Ashbel Hubbard* in July 1946. Even in 1947 when the first big consignment of poles (116,000) was for shipment, the major portion (70,000) was contracted on liner terms with the Head Bell Group (30,000) and Irish Shipping (40,000). The remainder was divided on a time charter basis between seven Finnish and Swedish ships, which between them contracted for thirteen voyages.

As Finland is a land of lakes, water was used to a great extent for the transport of poles from the forests down to the ports. On Lake Saimaa, north of the port of Hamina and close to the Russian border, large rafts made up of thousands of poles were regularly towed up to eighty miles on their way to the port. These usually started off as small rafts in the upper reaches of the lake, being added to as they progressed down the lake and finally finishing up with many thousands of poles. The final count of one of these rafts, destined for REO was 5,401 poles with a total volume of 62,125 cubic feet of timber. This was regarded locally as a world record.

Even with chartered shipping, all was not plain sailing. Delays were inevitable, owing to the uncertainty of loading and discharging times and time at sea. Further delays were caused by disputes at the loading ports, mostly regarding the time allowed for loading and the facilities offered, as some of these ports were tiny and very short of proper harbour installations. These delays in turn caused problems at the Irish end, as uncertainty of arrival dates made it very difficult to organise berthing and discharge facilities.

Once berthed, there was every incentive to unload and turn-around the ships as quickly as possible. The chartering agreements ('charter parties') included stipulations as to the time allowed for unloading and there was a heavy financial penalty, demurrage, payable for delaying the vessel's scheduled departure. Dublin port in the immediate post-war years was experiencing a boom. Container traffic was only in the early stages of development and the quaysides were congested with merchandise of every type awaiting customs clearance and collection. In order to speed up unloading operations a system was devised for directly discharging the REO poles on to transport vehicles which operated a shuttle service to the ESB's own pole-stacking areas which were well away from the congested quaysides. This required a high degree of organisation and synchronisation but it was so successful in achieving a high rate of discharge that it was adopted as standard at all three port depots.

Inevitably, ships sometimes arrived at congested ports and were unable to unload immediately, thereby putting REO in danger of incurring demurrage charges. Usually, however, extra effort on the part of the dockers recovered the position and over the total shipping campaign involving hundreds of voyages, demurrage payment was almost unknown. On the other hand, sizeable refunds were frequently obtained as the dockers, time and time again, discharged the ships well under the time allowed for in the charter party. In the case of one large ship, the *Maria Lisa Nurminen,* the standard discharge time allowed in the charter was eight working days. On two occasions the cargo of over 10,000 poles was discharged in five days and on a third occasion 10,460 poles were unloaded and stacked in four days. It is no surprise to find that the captain conveyed to the owners his preference for trading with Irish ports. Sometimes the speed of discharge caused a little embarrassment, as when the Master of the S.S. *Airisto,* discharging 11,326 poles at Limerick, had to alter plans to fit in both a trip to Killarney and the Dublin Horse Show during the allowed eight-day discharging period: he had to sacrifice Killarney and rush back to take his ship out three days earlier then he had anticipated. Limerick became a very popular port with the owners of that vessel.

In August 1947 while one ship was discharging its poles in Dublin a second pole ship arrived with no berthage space available — and none likely to become available for several days. With heavy demurrage charges looming, the problem was considered by the REO staff. One suggestion was to drop the poles over the side of the vessel into Alexandra Basin and construct a raft of the poles similar to those on the Finnish lakes. Information on these was available as, on one of his visits, Neil O'Donoghue had made meticulous sketches showing the construction of such rafts. The operation, however, was one hitherto outside Irish experience. The ESB often, of necessity, recruited young engineers fresh from college and threw them quickly in at the deep end. Here, at the very beginning of operations, the opportunity occurred to do this almost literally. The phone rang at headquarters. A young engineer was in the front hall reporting for duty. The reaction of the Engineer-in-Charge was immediate. He later recounted the incident: 'I dispatched my deputy, P. J. Dowling, to the front hall with explicit instructions. "Can you swim?", "Yes, fairly well", "Right! Go down immediately to Alexandra Basin and take charge of the construction of the pole raft." He complained to me later, "On my first day on rural electrification, I didn't even get past the front hall of the ESB. Instead I found myself hanging on grimly to the side of a ship contemplating the possibility of a watery grave as poles in their hundreds whistled past my head, to land with a smack in the murky waters of the Basin"'. In the event the operation was successful. A large raft of some thousands of poles, based on Neil O'Donoghue's sketches, was built in the best Scandinavian fashion by ESB pole-field staff. The methods used for slinging, tying and towing were all taken from O'Donoghue's notebook. The completed raft was towed down the Liffey by the Port and Docks tugs *Norway* and *Poolbeg,* brought around with some little difficulty into the estuary of the River Tolka and moored at the back of the pole-field, whence transfer to the stacking area was comparatively easy. The only really anxious moment was when it was necessary to 'put about' almost 180 degrees in order to get up the Tolka. An ebbing tide and a westerly wind conspired to send the raft across the Irish Sea in the direction of Liverpool. However, the planning had been well-done. The tide changed and a combination of flowing tide and herculean efforts on the part of the tugs persuaded the raft to change course and enter the calm waters of the Tolka.

By the middle of November 1947, of the 114,000 Scandinavian poles purchased, 95,000 had been delivered. Two ships had been chartered to lift the remaining 19,000 poles before the ice set in. Cables received on 20 and 24 November advised that both ships had 'touched ground' on their way to the loading ports. A later advice was that one, the *Scandnavic,* had broken up, fortunately without any loss of life. The 19,000 poles had consequently to be overwintered in Finland.

It was a tragic coincidence that on the very day in August 1947 that his sketches were being used to construct the ESB pole raft in Alexandra Basin, Neil O'Donoghue should have met a sudden and untimely death. He was followed in the job by Peter Conroy, an accountant who brought to the business of chartering a keen bargaining ability and an intense interest in all things nautical. Conroy was responsible for the shipment of the greater part of all the poles used on rural electrification — over one million. He revelled in the cut and thrust of offer and counter offer. He rapidly built up a first rate intelligence on the availability and going rate of the different ships in the pole trade and became recognised by shipowners in the Baltic trade as a client who was difficult to hoodwink. He was a good storyteller and many of the incidents quoted here are taken from his contributions to *REO News* and the *ESB Journal.*

One vessel deserves special mention, the venerable S.S. *Satakunta,* which year after year carried poles from Finnish ports to the Irish pole depots and became the best known ship of the pole fleet. A vessel of 2,207 gross tons, she was built in 1898 by W. Gray & Co. Ltd. at West Hartlepool and christened *Everest.* In 1939 she was sold to the Satakunman Laiva O/Y and registered under her new name at Pori, Finland. Her dimensions made her very suitable as a pole carrier. To quote from a letter to the editor of the *Evening Mail* on 13 October 1950 — '....she is, after over half a century of seagoing service, a great credit to the workmanship of the men who built her'.

Above: The heavily laden MV *Makè* had two inches to spare in crossing the bar to Limerick docks in July 1959. *Below*: Erecting the first pole at Kilsallaghan, Co. Dublin. On the extreme right is P. J. Dowling and next to him is W. F. Roe.

As more and larger vessels suitable for pole transport were located, and as the masters got more experience of the Irish ports and discharging performance, the amounts carried per ship increased, thereby keeping costs down.

Sometimes it was touch and go whether the large, heavily loaded vessels would clear the harbour bar. In August 1953, the *Airisto* was due in Limerick at the time of the August bank holiday. She was no stranger to Irish ports, having discharged 10,500 poles in Cork a couple of months earlier. This time she was carrying 11,326 poles and was fully down on her marks. Her estimated time of arrival would, however, coincide with neap-tide and it was feared that her draught would exceed the requirements of Limerick Port for this condition. The shipowners were alerted to the tidal conditions predicted, and it was suggested that if there were any doubts the vessel could be diverted to Cork. However her Master, Captain Ingerttila sent a message from Dover that the ship was now sailing on a lesser draught than on the earlier part of the voyage and that he would pick up the Limerick pilot at the mouth of the Shannon. The good weather would also be a help.

Arriving at Kerry Head the ship was on an even keel. Ballast tanks had been pumped, bunkers were at a minimum, and some of the crew even jested that drinking water was scarce. The vessel was now drawing 16′6″ — almost precisely the minimum clearance required by the harbour authorities for the expected tide. The pilot was taken aboard for the forty-mile trip up the estuary. The calm conditions held and at 6pm on August Monday the *Airisto* was safely berthed at Limerick Docks. Despite the absence of shoreside cranes (which necessitated the use of the ship's own winches), the ship was discharged in five days. Captain Ingerttila returned hurriedly from his planned tour of the southwest. His ship was ready to take out on Saturday afternoon instead of the following Tuesday as he had expected.

Even this virtuoso performance was overshadowed when, in July 1959, news came through that the M.V. *Makè* had scraped into Limerick Docks. The verb describing the vessel's arrival appears to have been well chosen, for, with a record cargo of 13,000 poles, the vessel was drawing 18′ of water. The clearance at high tide on that night when the vessel arrived was predicted as 18′2″. Normally a vessel can gain some extra clearance by emptying the ballast tanks. In the case of the *Makè* however, with over 50% of the cargo above decks, the emptying of the tanks would cause stability problems. Peter Conroy recalled that his main task on that occasion was to calm the fears of the harbourmaster until the vessel had finally come to rest at the quayside. The unloading of the 13,000 poles was completed in five days, which made some amends for the strain on the harbourmaster's nerves. Limerick Port was experiencing its busiest spell for over three years and berths were at a premium.

In September 1954 the S.S. *Karen* carrying 6,800 poles from Hamina to Limerick encountered hurricane conditions going through the Skagerrack. 2,400 of her poles were deck cargo. The vessel, a Danish ship, had been chartered for three voyages. On this third trip the trouble started as the ship left the Baltic, heading north through her home waters of the Kattegat. A few extracts from her log give some idea of the hazards of the sea, which are such an everyday reality to the mariner and so readily ignored or forgotten by the rest of us.

September 15 Left Copenhagen (after bunkering). Draught 14′8″ fwd. and 15′6″ aft. Fresh S.W. wind and passed the Skaw at midnight.

Sept. 16 Similar weather, heavy seas. Steaming only 5 knots and shipping heavy water over decks. Crew overhauling deck lashings. Wind increasing to gale force. Had to heave ship to.

Sept. 17 Wind WSW hurricane strength. Vessel hove to, shipping heavy water over cargo. At 6 am shipped heavy sea shifting deck cargo and some poles washed overboard. Deck lashings renewed. At 8 am a heavy sea caused further shifting of deck cargo. Vessel took heavy list to port and went athwart the sea losing steering way. A further heavy sea caused additional shifting of deck cargo and some more poles washed overboard. Port tank pumped out to reduce list and vessel put about. Proceeded to Fredrickshaven and moored there.

Over the next six days the deck cargo was unloaded, repairs carried out and deck cargo reloaded. The vessel bunkered and left Fredrickshaven on 23 September. Her troubles, however, were not yet over.

Sept. 24, 25, 26, 27 and 28 Fresh S.W. wind, strengthening S.W. wind. Gale force. Hove to. Vessel labouring heavily.

Sept. 29 Whole gale. Decks full. Vessel labouring heavily and decided to proceed under Duncansby Head for shelter.

Sept. 30 Weather improving, wind veered to east. Proceeded on passage and passed through Pentland Skerries.

Oct. 4 Arrived Shannon Estuary 6 am. Anchored awaiting pilot. Docked Limerick 5.30 pm.

The ship had taken twenty-five days to make the journey and had run the gauntlet of two storms of hurricane and whole gale force. It delivered 6,600 poles, enough for a further six rural areas. It is a fair assumption that the crews that erected these poles or the consumers who eventually enjoyed the electricity were completely unaware of the hardships and dangers undergone by those who transported them from the Baltic.

THE DONEGAL COASTERS

While large ships were busy carrying poles in bulk from the Baltic to the three Irish port depots, small coasters were at the same time transporting the complete package of materials for a number of rural areas from the same depots to a multitude of tiny ports, piers and jetties on the western and north-western seaboard, in most cases right up to the doorsteps of the areas to be serviced. Not that the coasters were left out of the Baltic runs. While their capacity was small – generally varying from about three thousand poles for the larger coasters down to one thousand or even five hundred on the small vessels – this in many cases was an advantage. It enabled them to slip into the smaller ports and lift small parcels of poles which would not warrant a larger vessel. Even when they had to make three or four voyages as against one voyage of a larger vessel, they were still competitive. One Dutch shipping company, Gruno Shipping, which operated a very large fleet of small coasters carried the bulk of the ESB pole shipments from the Baltic to Dublin, Cork and Limerick in the 1953 shipping season using two small ships, M.V. *Martien* and M.V. *Thérèse.*

In 1954, when the final cargo of the season had arrived in mid-December, twenty-four trips of about nine days in each direction had been made to lift 99,000 poles at five Finnish and two Swedish loading ports as far north as Toppila in the Gulf of Bothnia and as far east as Hamina on the Gulf of Finland, about twenty kilometres from the Russian border. There was one cargo of 10,460 poles and three of over 6,800 but the balance was carried by the small ships, averaging under 3,000 poles per trip.

Again in 1956 the *Invotra* fleet of coasters, also Dutch, shipped 60,000 poles – 40% of that year's quota – to the Irish ports. In that year a record total of 152,000 poles valued at over £700,000 was shipped from the Baltic; 60,500 to Dublin, 56,000 to Limerick and 35,000 to Cork.

The versatility of small coasters was further demonstrated when in the early summer of 1954 it was necessary to purchase six thousand creosoted poles from Sweden in order to keep up supplies of the more popular sizes until the creosoting of the new raw poles commenced in October.

Above: A skinning crew in Cork — they skinned the bark from the imported poles prior to creosoting. *Left*: Delivery of poles by coaster. *Below*: One horse power!

The first cargo of three thousand poles was loaded at Gothenburg for Dublin, both of which are large well-equipped ports. The second cargo, however, which was also of 3,000 poles was loaded on the Dutch coaster M.V. *Muphrid N* at the Swedish inland port of Otterbacken situated on Lake Vänern which meant the vessel had to be small enough to traverse the Gotha Canal. This cargo was brought direct to the ports of Galway and Westport, an achievement that would have been impossible with the larger ships.

The Electrification of County Donegal □ The story of the coasters is, however, so bound up with the story of the electrification of County Donegal that it would not be possible to recount one without the other.

Situated as it is in the north-western corner of Ireland, bounded to the west and north by the Atlantic Ocean and to the east and south east by the border with Northern Ireland, Donegal's only direct land connection with the rest of the Republic is through a very narrow throat at Ballyshannon in the extreme south of the county. All traffic from the Republic not passing through Northern Ireland must cross the River Erne at Ballyshannon bridge, then go north on the one road to Donegal Town. Here the road divides, one branch heading north towards Letterkenny, the other west towards Killybegs, Ardara and Glenties. Recently, with the help of EEC funds, the road has been considerably improved but in the late forties and early fifties it was narrow and twisting, most unsuitable for heavy commercial vehicles.

The physical isolation of the county from the rest of the State and the poor communications caused it to fall very much behind in infrastructural development. In so far as connection with the national electricity grid was concerned, Donegal had perforce been left behind in the large expansion of the thirties, when practically every town of any size in the rest of the State had got ESB supply. At the end of World War II, the only towns in County Donegal serviced by the ESB were Letterkenny, Lifford and Convoy, all fed from the Northern Ireland grid by means of a cross-border link at Strabane. The total number of electricity consumers so supplied was 750. All other towns and villages, some of considerable importance such as Ballyshannon, Bundoran, Buncrana, Donegal Town, Ballybofey, Stranorlar and Killybegs, were dependent on local electricity undertakings which, while mostly adequate to supply light domestic requirements, could not cater for the large motive power loads demanded by industry. The economic development of the county was therefore being held back, not alone by its geographical isolation and poor communications, but by an inadequate public electricity supply.

The difficulty was the remoteness of County Donegal from the existing grid. Sligo Town, 41 miles south of Donegal Town, was the nearest point of ESB supply, but at that time even supply to Sligo itself was not very secure, dependent as it was on a single 38kV 'tail' fed from the main grid station at Carrick-on-Shannon, 34 miles further away to the south-east. Without major and prohibitively expensive development the extension of ESB supply into most of Donegal did not appear feasible.

The breakthrough came with approval by the government in April 1945 of the hydro-electric development of the Lower River Erne with generating stations at Ballyshannon and Cliff which meant that a large and secure source of electricity would shortly be available within the county. The 38kV line was extended from Sligo to Ballyshannon as an advance link to service the construction sites while plans were developed to run heavier transmission lines to feeding points on the national grid.

Immediate advantage was taken of the connection with Sligo to change the town of Ballyshannon over to ESB supply and by March 1946 this had been completed. Supply was extended to the adjoining seaside resort of Bundoran in 1947. The government had given a directive that in order to get the Rural Electrification Scheme off the ground in as widespread a manner as possible, one rural area should be commenced in each county to start with. The arrival of the ESB supply at Ballyshannon made it feasible to comply with this directive. Rossnowlagh rural area adjacent to the town was commenced in August 1948.

With the completion in 1950 of the 123 miles of 110kV line from Ballyshannon to Dublin, Donegal's link with the national grid was secure. From Ballyshannon northwards, firstly to Donegal Town and thence through the Barnesmore Gap to Letterkenny, a 38kV trunk line was extended capable of meeting all immediate demands. By 1950, therefore, all was ready to commence the intensive electrification of the county.

In September 1950 work began on the area stretching west from Donegal Town to Mountcharles, thence on to Dunkineely, Killybegs and Kilcar. The north-east saw the crews arriving in the Raphoe area in September 1951. By 1955 four separate crews were busily erecting rural networks in different parts of the county and by the end of 1956 supply had been extended to twenty-eight rural areas with about twelve thousand new electricity consumers. By 1960 the 38kV line had been run in a complete loop around the county, increasing the security of supply immensely, forty-five areas had been completed and over twenty thousand new consumers connected.

In the very early stages of the development the weakness of the road communication between Donegal and the rest of the State was a problem.

The delivery of construction materials by road from the three main port depots of Dublin, Cork and Limerick was difficult enough even in the case of the rest of the country, hampered as REO was by the still serious shortage of heavy-duty road transport. The available fleets were very hard-pressed to meet the ever-increasing construction rates being attained in the field. In the case of Donegal, the position was near-critical. The road distances involved were immense. To Dungloe for example, the road mileage from the nearest pole depot, Dublin, was 213 miles, while from Limerick it was 220 miles. Nor was rail an alternative. Many of the Donegal areas then being selected for development had no rail facilities or only narrow-gauge lines, some of which were then in the process of being dismantled. There was also the problem of cross-border transport, always at prohibitive rates owing to the double and treble handling involved.

The experience gained in the use of small coasters in the Baltic trade now came into play. The possibility of using these to transport the large quantities of material required to areas on the west coast, particularly in Donegal, was examined. For a county with such a long coastline – four hundred miles, extending from the shores of Lough Foyle to the coast of County Leitrim – Donegal is not endowed with many large ports. It has, however, a large number of small harbours, jetties and piers, some of which were built as Famine Relief works or by the old Congested Districts Board, many of which had not seen even the smallest cargo boat since the CD Board had ceased its activities earlier in this century. There were difficulties in getting in and out and in most cases a vessel of any size would have to rest on her bottom at low tide.

There were, however, great advantages in cost and speed in the complete furnishing of an area with materials in one fell swoop, as against slow piecemeal delivery over the very poor roads of the time. REO was lucky in being in contact with coastal shipping companies which had wide experience of slipping in and out of tiny ports in all parts of Europe. One of these was the Dutch Gruno Shipping Company, which carried the bulk of the 1953 Baltic pole shipments. The story of the first coastal cargo shipped by this company is worth recounting in some detail. On Tuesday 6 October 1953, one of the company's vessels, the M.V. *Marwit* left Cork for the port of Ramelton in Lough Swilly with a cargo described in the ship's manifest as 'Materials in boxes, crates, coils, bundles and loose as per vouchers and 1,329 electric light poles'. The cargo was more elegantly and romantically described by a scribe of the time: 'The angels smiled because they knew that the cargo of the *Marwit* was nothing so prosaic. With that knowledge which is denied to men they knew that the little boat was laden with Christmas presents, with presents of electric light and power and the comfort and prosperity that goes with them for the good people of Ramelton and Rathmullen in County Donegal.'[1]

So anxious were all concerned to transport the maximum amount, that the vessel was right down on her marks, drawing 10′9″ of water. Captain Baarscheers, a veteran of countless forays into nominally inaccessible ports, was only slightly worried by the knowledge that the maximum clearance over the bar at Ramelton was officially advised as being only 10′ and that would be available only at high water of spring tide due at 6.00 pm on the following Thursday. He thus had two problems: firstly to arrive at the harbour bar not later than high water, and secondly to nurse his ship over the bar. Proceeding via the Irish Sea the vessel rounded Malin Head into Lough Swilly in good time to pick up the pilot at Inch Island. To quote from the contemporary report: 'Pilot Brown held her back to a steady two knots for the last three miles of her journey as the spring tide rose beneath her. Precisely at 5.45 pm she crossed the bar without incident and tied up at Ramelton quay where about 500 people had assembled to welcome her.'[2]

The unloading was carried out immediately and speedily using the ship's own winches operated by the ship's engineer and first mate, neither of whom spoke English, assisted by an unloading gang of local men who knew no Dutch. However, with shouts and internationally understood signs the unloading was completed speedily if somewhat noisily with just one minor casualty as the 6′7″ ESB chargehand Jack Newton was knocked into the water by a sling of poles to the loud cheers of the audience on the quayside. As the local Garda Sergeant expressed it 'There hasn't been such excitement in Ramelton since the Civil War!'

The next morning the *Marwit* again crossed the bar in the morning mist, this time en route for her next contract in the Algerian sunshine. A month later the M.V. *Martha,* also of the Gruno line, arrived in Ramelton with a second load. These operations were somewhat unusual in shipping circles as the ESB were suppliers of cargo, charterers, receivers, port agents, brokers, ships' agents and stevedores, all at the same time.

In the next few years practically every usable harbour, jetty and pier on the Donegal seaboard and even a private landing stage at the Capuchin monastery on the Ards Peninsula near Creeslough were used to land cargoes. As most of the piers were prohibited landing places in closed ports, except to fishermen, the permission of the Revenue Commissioners and the approval of the Department of Industry and Commerce had to be obtained in advance. For each shipment, the cargoes were loaded mainly at one of the three port depots and occasionally at Galway, Sligo, Ballina and Westport, where some overseas consignments had been discharged. On one occasion a five-port, seven-point operation was carried out by the M.V. *Flevo,* loading at Dublin, discharging and reloading at Cork, discharging at Galway, Cleggan and Newport, and returning to Cork to load a cargo for Dublin. It is sad to record that this ship was lost with all hands in the North Sea a few years later when it was not, however, on REO charter.

The coasters used were not all Dutch, of course. The Limerick Steamship Company vessel M.V. *Galtee* was constantly engaged in coastal deliveries. One voyage of this vessel worth recording took place in October 1954. Actually there were two ships involved, the second being another Irish ship, the M.V. *George Emilie* of Arklow. The *George Emilie* loaded at Dublin for a two-port discharge, first at Bunbeg and then at Letterkenny. The *Galtee* loaded at Cork for a three-port discharge — Dublin, Letterkenny and Burtonport.

Letterkenny port at the time had only one berth and was at the head of the narrow five-mile navigation of the Swilly River with its outlet into the southernmost reach of Lough Swilly. The timing called for the *George Emilie* to have discharged at Letterkenny and have cleared the Swilly River in time to allow the *Galtee* to enter with her cargo. Despite the most severe storms of the year, both ships kept to time and the discharged *George Emilie* cleared the mouth of the river on schedule, passing on its way down the Lough the loaded-down *Galtee* at its moorings awaiting the rising tide. However, just as the Agent and the ESB staff had collected the local pilot for the river and set out from Letterkenny for the anchorage, puffs of smoke were seen in the distance ascending the horizon in line with the ascending tide in the river. To quote from the contemporary report:

> As we drove along the bank of the river towards the Lough, a small steamboat flying the Union Jack was observed in the mouth of the river, the funnel spitting and spluttering steam and smoke in an effort to keep the vessel afloat as there was not yet enough water in the river to keep the heavily laden boat off bottom. The skipper had evidently decided to enter the river while daylight lasted and inch his way up on the rising tide with his cargo of coal. It was a small vessel and could make port on a draught several feet less than the *Galtee.* As there is only one berth in Letterkenny, a midnight conference was held aboard the *Galtee* after a perilous voyage to the anchorage in a rowing boat. It was decided to miss Letterkenny and sail for Ramelton where discharge was completed on to about 60 feet of pier, the remaining quayside having fallen into the water sometime earlier in the year. The *Galtee* flying the tricolour was the biggest ship ever to go up the River Lennon from Lough Swilly to Ramelton. Discharge at the

> third port, Burtonport, was uneventful but on leaving, the *Galtee* ran into a continuation of the severe gales and had to shelter in Killybegs for several days.[3]

In contrast to these conditions were some balmy days in April 1955 when the *Galtee* again arrived in Lough Swilly bound for Ramelton with a cargo of poles for the Portsalon area some fifteen miles to the north. Because conditions were so perfect the captain decided to land the poles on to Portsalon pier right in the heart of the rural area where they were required. This pier is situated only five miles from the mouth of Lough Swilly on the Fanad peninsula and, being subject to the prevalent Atlantic swells, was hardly ever used. The weather held long enough to enable full discharge,

> much to the disappointment of the people of Ramelton, 15 miles down the Lough who were awaiting the arrival of the ship as scheduled. However, when the ship returned a week later, this time for a scheduled discharge at Portsalon, heavy swells prevented this and the ship had to proceed to Ramelton where the people ... turned out in force on Low Sunday evening to give the *Galtee* a rousing welcome and ... an assurance to the Captain that they were pressing the County Council to have the pier repairs completed before his next visit.[4]

Not all shipping was to such sheltered waters as Lough Swilly. At Teelin, Downings, Burtonport, Bunbeg and Malin Head, even at Donegal Pier itself, the turbulent Atlantic frequently made itself felt. When the island of Arranmore came up for development, delivery of the required poles and other materials presented a problem as there was no suitable pier for the discharge of even a small coaster and the prospect of ferrying materials out piecemeal in the half-deckers that normally serviced the island was a daunting one, both in regard to time and to cost. Finally, 'after much persuasion of the shipowners' it was agreed to beach the Dutch M.V. *Whitsun* loaded with five hundred poles, cable, transformers and other equipment, including a delivery van and three compressor-tractors, on a selected shallow-water section of the fore-shore. Close to the beaching site were several small, partly covered stags of rocks, very dangerous in stormy weather. The ship was beached on the island on Sunday afternoon, and when the tide receded at 8.00 am on Monday morning unloading began. Everything went well until the evening of the third day when a north-easterly gale hit the area, raised a heavy swell on the spring tide running at the time, dragged the anchor, smashed the moorings and almost tossed the ship onto the island only fifty yards off and then lifted it within a few feet of the rock. Only the skill and heroism of the ship's captain and the Arranmore lifeboat crew saved the ship from disaster. At the height of the storm the captain's wife and child were taken off. The *Derry Journal* ran a leading article in its issue of 11 April entitled 'Hats off to them'.

Some captains were not as adventurous. At Portsalon the captain of one Irish coaster refused to make a return trip when the locals told him that a coaster of the size of his vessel had not been seen there for more than thirty-six years. The narrow pool of tidal water which served as the port of Letterkenny on the River Swilly was also refused by several owners after a few exciting trips there with quite small boats. For the record, navigation conditions have been improved immensely since then, allowing much larger vessels to ply regularly to the port.

Away from the exposed Atlantic coast, in the relatively calm waters of Lough Foyle, battle was joined between two vessels for the first occupancy of a single berth pier. The story can best be recounted in the words of Peter Conroy who as honorary Commodore of the REO coaster fleet could not be regarded as a completely neutral observer:

> In the earlier case, we had a race for a single berth port [Letterkenny] between our chartered ship, the M.V. *Galtee* and the S.S. *Mulcair* carrying coal out of a British port. We lost that race, perhaps because we didn't know all the tricks of the trade, but we learned something from the losing and on Friday, 26th August, of this year [1955], we found ourselves once again in competition. This time our candidate was the Dutch motor vessel *Port Talbot* (350 tons) and the opposition was the S.S. *Ronaic* with 500 tons of coal from Maryport. The goal was the single berth pier at Moville, known locally as the 'Famine Pier'.
>
> It was a hot, rather oppressive day that 26th August and large crowds of holidaymakers and daytrippers were assembled at vantage points on the western shores of Lough Foyle. The main attraction from their point of view was the approaching exercise of the NATO forces due to take place in the Lough. The local people went about their tasks in their usual calm mood but the righteous strained their eyes towards the mist-enshrouded Magilligan Point and Innishowen Head at the mouth of the Lough, where the big battle craft would appear as soon as visibility permitted. A small handful of people also watched anxiously towards the points, and the question uppermost in their minds was would the *Port Talbot* make it before *Ronaic* because our land forces were mobilized at Moville ready to unload the poles and other materials for the construction of Muff Rural Area.
>
> Under the heat of the August sun the mist lifted, and round the headlands came the first of the great ships. A stir of excitement ran through the watching crowds as Portuguese, Netherlands and Norwegian naval units

moved into the Lough to join the British in occupation there. The full panoply of war was apparent: air and submarine escorts fussed hither and thither, all anxious to show their paces and join in the mock battle further down the Lough.

As the mist lifted further and the battle fleets stretched themselves out after the narrow entry into the Lough, the tenseness among the small handful watching the lesser struggle increased: for there surely, sandwiched in among their bigger sisters, were the two coasters, dwarfed down by comparison to the size of fishing smacks.

There was a general heave to as pilots came aboard, and then the watching few noticed that the *Port Talbot* seemed to be doing a lot of apparently aimless dodging about, but the more shrewd observers knew that this was not aimless; this was tactics and the race would be to the tactician. Seamanship was at a premium now, and the Dutch captains of coastal vessels are masters of seamanship. The expert observers relaxed their tension and even the landlubbers could see the pattern of manoeuvre. Suddenly the Dutch vessel was away to a quick start and was well in the navigation channel. It was almost like a good coursing greyhound getting away quickly from the slips, and the first 'turn' and the initiative were with the *Port Talbot.* The small group of spectators exchanged quiet grins: the race was in hands, and in the cool of the August evening the *Port Talbot* was at Moville Pier, and the *Ronaic,* ploughing through its stern wave, must lie 'off'.

The big battleships sailed on down the Lough and the greater battle was joined but at Moville Pier the land crews were already discharging the *Port Talbot* to the accompaniment of zooming aircraft, the fire of heavy guns, and the brilliant light of target flares and markers of the NATO exercise.

The captain of the *Ronaic* looked a little disappointed when he came ashore from his anchored position off the pier. He had been beaten in the race, and, of course, no one particularly likes that, especially when time and money are involved. He seemed surprised that unloading had commenced at that late hour (it was now 5pm) and would finish next day. Such a thing, he said, could only happen in the 'Free State'.

Late the following night when the *Ronaio* had moved into the vacated berth and the *Port Talbot* was ready to sail, the two captains forgathered with our officials in the local hotel and the story was told of the earlier race we had lost. The toast was 'revenge for the Swilly'.[5]

While Donegal tended to occupy the centre of the coastal shipping stage, down along the west coast the little ships sought out the small harbours bringing with them a new way of life in the form of poles, transformers and drums of cable and indeed wakening up many of these ports from a long slumber. The list is almost endless: Malin Head, Moville, Buncrana, Culdaff, Letterkenny, Ramelton, Portsalon, Downings, Ards, Bunbeg, Burtonport, Teelin and Donegal Pier. Further down the coast was the port of Sligo and, in County Mayo, Westport, Newport, Ballina and Belmullet. In Belmullet the M.V. *Whitsun* arrived with poles for Geesala rural area and provided considerable excitement for the community, particularly the children who were seeing the first ever 'big' ship in their port. In the current progress report for the area it was solemnly recorded that 'during the discharge, the captain received numerous local businessmen and Church dignitaries aboard the vessel.'

Cleggan in County Galway, Portmagee and Valentia Island in County Kerry, Castletownbere and Beare Island in Bantry Bay also saw the coasters delivering not only the materials but the machinery and transport required to carry out the works. The M.V. *Whitsun* had to be partially beached alongside the stone jetty at Beare Island to unload its cargo which included three tractors.

All in all, the coastwise shipping operation was a success. It made it possible to construct electricity networks even in remote areas to reasonable time and cost targets by delivering the materials in bulk right on the spot where and when needed. The enthusiasm and commitment shown by staff involved, both of the ESB and the shipping companies, was remarkable. Calculated risks were taken by many captains in approaching ports without adequate, up-to-date charts. Many of the ports had long been silted up and were fringed with foul ground of one kind or another, not to mention the always turbulent Atlantic ocean. The REO land crews, usually of local young men, engaged in the loading and unloading operations, appeared to be almost obsessed with demonstrating how quickly they could load and discharge and they would obviously take it as a defeat if a vessel missed a tide because of delay on their part. Despite the apparent frenzy of the loading and unloading, only one accident (not fatal) was recorded against the activity on its five years of operation.

CHAPTER NINE

'Sing the Peasantry and then . . .' The People — 1946

THE POPULATION TO BE SERVED BY THE SCHEME was about 1.75 million, of which about 35,000 (or about 2 per cent) lived in small villages. The remaining 98% lived in the open countryside, sometimes in small clusters of houses, but generally in scattered dwellings. It was estimated that about 80% of the total target population depended on farming for a livelihood. The remaining 20% was made up of rural workers not in agriculture, town workers living in the country, government officials, shopkeepers, publicans, pensioners, clergy, schoolteachers. To this non-farming group the coming of electricity meant an opportunity not so much to increase productivity as to improve their amenities and standard of living. Generally they had comparatively steady incomes. Many were already familiar with the benefits of electricity and were more than anxious to avail of these without delay. It was from the ranks of these non-farming rural dwellers as well as from some of the more progressive farmers that the instigators and organisers of the local rural electrification committees very often emerged.

The success or failure of the scheme, however, would depend on how it was received by the 80% of the rural population who depended on farming for their livelihood. From the very conception of the Shannon Scheme, its eventual extension to include the farms of Ireland was envisaged by its planners. It was realised that the prime objective – the transformation of the new State into a modern socially and economically progressive community – would be only partially achieved without the inclusion of the farming sector. Unless the quality of farm life was improved and the low productivity of the majority of farms lifted, the main export from the country would continue to be people, mostly the young and active and usually from a farming background.

The experience of the US, Canada and many countries of western Europe over the thirties confirmed that the coming of electricity to rural areas had sparked off a transformation of attitudes, living standards and economic productivity. It was realised, however, that the problems of rural electrification in Ireland would be unique in many aspects and that the approaches that were successful in other countries would not necessarily be so in the Irish context.

To begin with, the time of launching the scheme – immediately after World War II – was hardly propitious. Even in times of prosperity, the initiation of such a large-scale project would have been an ambitious venture. In the Ireland of the late 1940's and early 1950's with its depressed and slow moving economy, it was an immense act of faith.

In 1946, Éire, formerly the Irish Free State and later, in 1949, to become the Republic of Ireland, was still a very young political entity. It had had a difficult birth in 1922 and in the intervening twenty-four years had been buffeted by a series of economic and political storms which had left its economy in a depressed condition at the end of World War II. An agricultural country with a home market of less than three million people, it had traditionally depended on exports of agricultural produce, almost entirely to Britain. In the decade immediately preceding independence, which included the period of World War I, there had been an unlimited market in Britain for Irish produce at very high prices. Because of the war, competition from other suppliers was almost non-existent, and unfortunately too many Irish exporters took advantage of this to ship food of very poor quality. Shortly after the end of the war the bonanza ceased and prices collapsed. To quote one writer, 'When the war was over and supplies from Denmark and overseas were again available, the British consumer had only too clear a memory of bad eggs and worse butter from Ireland.'[1]

The prospect facing the government of the new State was a daunting one. On top of the collapse of the agricultural export market, the country had been badly shattered by the conflicts of the War of Independence and was to endure ever more from the succeeding Civil War and the period of severe civil unrest which followed. By 1924, however, the task of rebuilding was well under way. One of the first aims of the new National Government was to recover the export markets which had been lost. Regulations were swiftly introduced to ensure standards of cleanliness and purity in the packaging and marketing of pork, bacon, eggs, and dairy produce to such effect that by 1929 the quality image of these products had been restored on the British market and prices paid had risen to the level of the competitors. Penetration of other markets was, however, an almost impossible task and was to remain so until Ireland was admitted to the EEC in the early seventies.

1929 saw the beginning of the great world depression with its consequent detrimental effect on world trade. In Ireland this was followed by

One of the back-breaking chores of the farm wife in pre-electrification days.

the 'economic war' with Great Britain in 1932 when penal duties on Irish produce were imposed by Britain in order to recover the amount of the land annuities being withheld by the de Valera government. Yet again the British market for Irish agricultural produce collapsed. The total value of exports dropped from £36 million in 1931 to £19 million in 1933 and cattle exports from £18 million to £9 million. Ireland retaliated with duties on British goods. These stimulated the small industrial sector and employment in industry and services rose to meet the limited home market. This, however, was offset by the fall in the income of the farming sector as their costs rose and markets and prices toppled. In 1935 with the 'coal-cattle pact', and in 1938 the Anglo-Irish agreement, there was a recovery in quantity and price of agricultural produce on the British market, but this was not sufficient to compensate Irish farmers for losses in the early part of the decade.

In 1939 World War II broke out. This time there were not the bonanza prices of 1914–18 as a very deliberate cheap food policy and strong control by the British government held down prices in this, the only export market available to the Irish farmer. British and Northern Ireland farmers were encouraged towards maximum food production as part of the war effort and were compensated for low prices by heavy government subsidies. No such subsidies were available to farmers in neutral Éire. As the war progresed exports dropped to half the pre-war level while imports plunged to less than one-third of the 1938 figure. Imports of fertiliser and animal feeding stuffs and of wheat for flour milling dried up almost completely so that the country was compelled to feed itself from its own resources. Compulsory tillage was introduced and the area devoted to growing wheat rose to almost three times the pre-war figure. The country survived on a diet of home-produced meat, potatoes, eggs, butter and bread milled from home-grown wheat, which, while monotonous, ensured that it was still one of the best fed communities in the world during this difficult period.

By the end of the war the Irish economy, especially that of the farming sector, had suffered immeasurably. Import prices had risen by 122% between 1939 and 1945 while export prices only rose by 89%. As a compensation there was a large rise in foreign assets. This rise was not, however, the result of booming exports as in the 1914–18 period, but of falling imports as both foreign commodities and shipping became unavailable. Agricultural production stagnated as, in the absence of fertilisers and animal feeding stuffs, output was only maintained by drawing down the fertility of the soil.

An examination of the structure of the post-war farming in Ireland will also help in understanding the particular problems that made the task of rural electrification in Ireland in 1946 so challenging. The total acreage available for farming in the state (excluding woods, mountain, bogs, marshes etc.) is about twelve million acres of which at the time about seven and a half million or 60% were in pasture, 2 million devoted to hay and two and a half million (22%) in tillage. The number of agricultural holdings of over one acre amounted to about 320,000, giving an average size of holding of thirty-seven acres. However, there was a very wide variation from average both as to farm size and quality of land. Table 2 quoted in the 1944 'Report on Rural Electrification', shows the counties in which most of the acreage is in large holdings (predominantly the eastern countries), in medium-sized holdings (predominantly the southern counties) and in small holdings, the western group. Table 3 gives the break-down of farm size in the same three groups of counties. The large number of small holdings even in the first two groups is noteworthy. In the eastern group, 70% of holdings were under fifty acres, while in the southern group the figure was 62%. Thus, even in the most fertile parts of the country the majority of the farmers were of comparatively modest means. In England, by contrast, the average size of farm was about twice that size.

When one turns to the western group however, the gap between the two 'Irelands' is clearly seen. Historical, political, geographical, and geological influences had resulted not only in smaller farms (88% under fifty acres and 68% under thirty acres in 1946) but in poorer land, a less favourable climate and consequently lower average farm incomes than in the remainder of the country.

A survey of twenty farms of sizes ranging from ten to one hundred acres carried out in mid-Roscommon in 1945–46 by R. O'Connor gives a very good picture of the typical western dry-stock farm of the period:

> Meadow and pasture occupied 80% of the available acreage while tillage took up 18%, the remainder being unproductive. The average value of total output per farm was £416 which included a figure of £94 as value of the produce consumed in the home. This latter included milk, butter, buttermilk, bacon, poultry, crops and turf.
> Expenses excluding labour amounted to £96 leaving an average total labour income of £320 or 77% of total output. When the cost of hired labour per farm of £46 was deducted, the total family income per farm was £274 or just £5 per week. The survey showed that the average total 'family labour units' per farm was 1.87 (one labour unit represents the equivalent of an adult worker fully engaged for 52 weeks), giving an average annual family labour income per unit of £146.8 or £2.16.0 per week.

TABLE 2

Size-Distribution of Farms in the Various Counties

Aggregate agricultural acreage in each size-group as a percentage of the total agricultural acreage of each county at 1 June 1931.

(Arranged from Table 61, Statistical Abstract 1940, Dept of Industry and Commerce).

Group	*County*	*Up to 30 Acres*	*30-50 acres*	*50-100 acres*	*100-200 acres*	*Over 200 acres*
		%	%	%	%	%
Counties in which most of the acreage is in large farms.	Kildare	9	8	16	26	41
	Meath	13	10	14	22	41
	Dublin	15	10	20	26	29
	Wicklow	10	10	24	28	28
	Westmeath	16	15	21	20	28
	Offaly	14	15	25	21	25
	Louth	26	13	20	18	23
	Average	14	12	19	23	32
Counties in which most of the acreage is in medium sized farms	Kerry	19	21	35	18	7
	Limerick	13	17	34	25	11
	Cork	10	14	34	29	13
	Wexford	11	14	31	28	16
	Clare	19	22	30	17	12
	Kilkenny	10	14	30	28	18
	Tipperary	11	15	29	25	20
	Carlow	12	13	29	27	19
	Waterford	8	9	26	33	24
	Laois	15	14	25	26	20
	Average	13	16	31	25	15
Counties in which most of the acreage is in small farms	Mayo	53	22	12	6	7
	Leitrim	50	26	17	5	2
	Monaghan	46	24	19	7	4
	Sligo	45	22	16	9	8
	Cavan	44	26	19	7	4
	Roscommon	41	23	16	10	10
	Donegal	35	20	23	14	8
	Longford	34	23	21	12	10
	Galway	30	26	21	11	12
	Average	41	24	18	9	8

TABLE 3

Number of Agricultural Holdings Exceeding One Acre

1 June 1945

(Arranged from Table 61, Statistical Abstract 1946, Dept of Industry and Commerce)

	Up to 30 acres	*30-50 acres*	*50-100 acres*	*100-200 acres*	*200 acres and over*	*Total*
Kildare	3,125	778	823	692	417	5,835
Meath	4,437	1,791	1,270	867	542	8,907
Dublin	3,063	477	506	346	144	4,563
Wicklow	2,028	818	1,200	711	280	5,037
Westmeath	3,831	1,585	1,230	554	336	7,536
Offaly	3,278	1,576	1,466	663	294	7,277
Louth	3,493	713	538	252	114	5,110
Total	23,255	7,738	7,033	4,085	2,127	44,238
	(52.5%)	(17.5%)	(16.0%)	(9%)	(5%)	
Kerry	8,814	4,041	3,987	1,327	440	18,609
Limerick	5,125	2,676	3,080	1,220	244	12,345
Cork	9,726	5,513	7,731	3,846	821	27,637
Wexford	3,349	1,728	2,296	1,256	314	8,943
Clare	6,253	3,688	3,128	963	361	14,393
Kilkenny	2,696	1,546	2,010	1,119	338	7,709
Tipperary	6,562	3,313	3,859	1,825	578	16,137
Carlow	1,490	614	855	440	133	3,532
Waterford	2,107	830	1,339	955	314	5,545
Laois	3,318	1,322	1,346	718	248	6,952
Total	49,440	25,271	29,631	13,669	3,791	121,802
	(41%)	(21%)	(24%)	(11%)	(3%)	
Mayo	23,824	4,520	1,678	388	219	30,629
Leitrim	7,950	2,210	842	156	24	11,182
Monaghan	7,770	2,024	1,002	180	29	11,005
Sligo	8,659	2,205	935	242	98	12,139
Cavan	9,910	3,015	1,400	239	50	14,614
Roscommon	11,253	3,740	1,456	391	98	16,938
Donegal	16,286	3,465	2,638	1,055	384	23,828
Longford	4,040	1,366	750	228	79	6,463
Galway	15,859	7,095	3,287	808	347	27,397
Total	105,551	29,640	13,988	3,687	1,328	154,194
	(68%)	(20%)	(9%)	(2%)	(1%)	
Total (All counties)	178,246	62,649	50,652	21,441	7,246	320,234
	(56%)	(20%)	(16%)	(6%)	(2%)	

> If the value of the produce consumed in the farmers' homes was excluded, the annual cash remuneration per unit of family labour would be £69 on the under 50 acre farms; £128 on the over 50 acre farms, and £96 on all farms. Calculated at a weekly rate this would amount to 27/- per unit and the small farms (under 50 acres); 49/- on larger farms and 37/- on all farms.[2]

240,000 of the 320,000 holdings in the country at the time were of fifty acres or less — indeed 178,000 of them were thirty acres or less (see Table 3). It should be noted that the land in the part of County Roscommon covered by the O'Connor survey is by no means the least fertile and if this survey were taken as indicative of the situation of the small farmer in the country as a whole, then a very large proportion of the potential consumers of the Rural Electrification Scheme were involved in 1946 in a subsistence type of economy.

Dr J. J. Scully in a study carried out much later on farming in the west of Ireland contrasts this farming for subsistence with commercial farming. He defined the commercial farm as market-orientated, its output influenced not alone by the relative productivity of inputs but also by the relative profitability of alternative farm products and farming systems. Commercial farms by this definition there were in plenty, mostly in the eastern and southern parts of the country and mostly in the larger acreages. The owners of these farms, geared as they were to market demands, welcomed electricity as a means of increasing efficiency and productivity. In the low-income or subsistence farm on the other hand

> farm operations are usually performed in a traditional manner with the same inputs being combined in the same way to produce the same products from one year to the next... .
> The low income segment is normally static or maybe contracting. It is the segment which suffers the greatest depletion in manpower because of migration to other sectors with the progress of economic growth.[3]

This was the case in the majority of farms in the poorer districts, but Scully's description also applied to a very large number in the more prosperous parts. It was with these householders that the local rural electrification committees and ESB Area Organisers had to work hardest to persuade them of the benefits electricity could bring.

There were many factors inhibiting change. Clearly, many small-farm families were living in near poverty. Following the failure of the potato crop and the Great Famine of a century earlier, Irish farming had moved away from the intensive tillage systems and towards a heavy dependence on livestock rearing. The switch to low-intensity cattle production was understandable in the context of a depleted population on the land and poor markets for other produce, but this was not the kind of enterprise to generate high farm incomes and indeed it was badly suited to the small farm, particularly on the poor soils of the west.

There were also sociological obstacles. The Land War was still fresh in the folk memory. Land ownership rights had been dearly won and in too many cases security of possession was regarded as more important than progressive agricultural development. At the 1951 census 73% of the country's 199,000 male farmers were over forty-five years of age and almost a quarter of these were unmarried. Even where there were potential heirs working on the farm there was a general reluctance to yield decision-making or hand over ownership to the younger person. Educational levels were low: less than 5% of farmers had gone beyond primary school level. Agricultural advisory services were still comparatively underdeveloped. Perhaps most importantly the great rural movements, such as Macra na Feirme, Macra na Tuaithe, Muintir na Tíre and the Irish Countrywomen's Association, which were later to do so much for adult education and for social and economic development had not yet emerged to their full strength.

With this rather depressing combination of circumstances rural emigration was inevitable. Sons and daughters sought a better life elsewhere, leaving, in many cases, the least resourceful and least educated to run the farm — a tragic merry-go-round which almost ensured that the pattern would be repeated from generation to generation. The hard-won political independence of 1922 had not in itself, as many of its founders had hoped and expected, put an end to this delibitating haemorrhage of emigration which, generation after generation for over a century, had been draining the countryside, mostly from the farming community and especially from the areas of poor land and small farms.

By 1948 the government had become so concerned that it set up a Commission on Emigration and Other Population Problems to study the situation and make recommendations. The Commission identified the stagnation in agricultural production and the decline in agricultural employment as one of the prime causes of emigration. It reported that it would be necessary to increase agricultural output by fifty per cent in order to maintain the existing population on the land, but highlighted the immensity of this task by pointing out that there had been no growth in the volume of this output over the previous half century. If the system of agriculture were based on the intensive use of land, the Commission believed that not only could the agricultural population be maintained at its then level, but that it could be increased.

Turning to living conditions the Commission stated:

> In rural areas, many houses are quite unsuitable as regards design, accommodation and comfort, if judged by modern urban standards. The contrast between rural and modern urban housing is so marked that it must have a considerable effect on the decision of a person in a rural area whether to emigrate or to marry and settle down in his locality...
> Poor material standards of life are to be found in many parts of the country, but particularly in rural areas where the exacting demands of agricultural activities are aggravated, in a great many cases by the inadequacy of such amenities for houses and farms as power, light, water and sanitation. Much drudgery is caused by out-moded methods of day to day working and living.[4]

With regard to the Rural Electrification Scheme, the Commission observed:

> The national scheme of rural electrification should, in due course, considerably affect the lives of those in rural areas. The provision of electricity for power, light, heating and cooking, not alone in dwelling houses, but also in out-offices and farmyards should revolutionise many habits of rural life and in particular improve the lot of the housewife. It should enable many farmers to introduce improvements and modern amenities comparable with those of urban centres. It could ease the way for the return of the craftsman to rural areas, providing employment for him both in agriculture and in cottage industries.

Despite such endorsements of the scheme there remained considerable inertia to be overcome. Many farmers, particularly those outside the dairying areas,[5] with small acreages and uncertain incomes remained to be convinced that rural electrification could ease their lot. Not all could see in electricity the promised release from the many time-consuming tasks and the consequent opportunities to devote extra time to more productive activities. Indeed it was questionable if, at the time, profitable markets could be secured for such increased production. Not all husbands could see a necessity to ease the traditional drudgery of the farm housewife. To many, electricity first appeared merely as an expensive, if admittedly a greatly superior, alternative to the traditional oil lamp or candle. Their forebears had successfully survived without it and they could see no great advantage in hurrying to involve themselves in this new expense.

Above: The electricity arrives at Rosses Point, Co. Sligo, November, 1947. *Below*: The new electric pump — a source of wonder and of benefit to all in the home and on the farm.

That there was good reason for this cautious approach and that it was not confined to electricity was pointed up by Professor James Meenan in his minority report as member of the Commission on Emigration and Other Population Problems:

> The average age of male farmers was 55 years in 1946. The average farmer would thus have been 23 years old when the First World War broke out in 1914. Since then, he has lived through two world wars, the Anglo-Irish War and a Civil War; two collapses in prices including a major depression; four devaluations of currency, the 'Economic War' and all the business of reconstruction after 1918, 1922, 1938 and 1945.
> It is not perhaps a matter for wonder if farmers to-day take short views and are reluctant to embark on improvements that, however excellent they may be in theory, are only hostages to the fortunes of prices and to the political and economic events that control prices (p.377).

It is necessary to stress that the reluctance to change on the part of so many was based on more than suspicion of innovation or of the motives of the innovators. There was a positive side — holding on to a traditional way of community life that had been shaped by history and tempered in the fires of many holocausts. It was simple, sometimes austere, but basically wholesome. Hard physical work, neighbourly co-operation and a strong, simple religious faith were its characteristics. There was a symbolism in the fire on the hearth which was never allowed to go out. The lack of mobility outside one's own area contributed to a strong consciousness of community and the sense of belonging.

Despite the absence of wealth, it could be said that these poor rural communities were better adjusted to their straitened circumstances than their more affluent successors were to theirs. Incomes were low, but so were expectations. To quote John Healy: 'The rhythm was slow and fixed; the wants were small and if you didn't have it you went without. In those days you wanted for little enough on small holdings like these... .'[6] The more significant aspects of the cash economy had not yet permeated people's lives, especially outside the dairying regions where regular cheques for farm produce were not in existence. In a sense, the coming of rural electrification with its fixed charges and regular billing was a definite wedge in the change to a more money-conscious rural society.

In this context, the experience of Jim Wolahan, Rural Area Clerk in Dundalk District in 1955, is worth recounting as he told it to the writer.

> In July [1955] I was transferred from Oldcastle to Killanny rural area on the Louth-Monaghan border. As I had only a bicycle for transport I

enquired locally for a digs near the Area Office at Essexford. I was directed to a cottage occupied by the man of the house, his wife, two children and a 'granny'. It was spotlessly clean and they were willing to take me. When it came to fixing terms, the woman was nonplussed. She had never kept a lodger before. What had I been paying in Oldcastle? £3 per week. Well Oldcastle had amenities not available here — no toilet and as yet no electricity — how about 50/- or say 7/- per day. If I went home for week-ends she would only charge 35/- for that week. Later she decided that even this was too much. She pointed out that she had her own eggs and brown bread and apart from buying meat for my dinner had very little expense. Despite my protestations, she later decided that 25/- was sufficient as it transpired that I was now going home every week-end.
When the time came to transfer to a new area in south Monaghan, about 15 miles away, my personal transport had progressed from the bicycle to an old VW 'Beetle'. I was asked if I would remain on in the house and 'commute' to work and I agreed. However, the woman of the house pointed out that as I now had my lunch out, she would not have to buy any more meat for me. Her expenses were now almost nil and she proposed 15/- per week as my new rate. After some argument she compromised at £1!
One day granny announced 'I am going to look after your shoes'. She wanted to have some little function in my welfare which did not cut across the role of the younger woman. Every morning I found my shoes shining. There was not much place for money in that house.

Change was undoubtedly necessary if the rural economy was not to stagnate and its population dwindle to crisis point. Most rural people were aware of this, but the kind of change, its extent and its 'knock-on' effects gave cause for deliberation. Indeed writers such as Hugh Brody[7] would hold that, in the event, many of the recent social and economic changes have left these communities all the poorer. It must be left to future social historians to debate how much of a certain quality of life was lost with the efforts to eliminate the harsher elements of rural living in Ireland.

CHAPTER TEN

Moving In

IN THE EARLY STUDIES on the organisation of the scheme it was recommended that the rural parish, with an average area of twenty-five to thirty square miles and containing five hundred premises, would be a suitable geographical unit on which to base a rural electrification area. There were about eight hundred rural parishes in the country, not all, unfortunately, conforming to the average in either size or population. To quote W. F. Roe, 'the original parishes were formed by monks sent out from the monasteries and it depended on the energy or stupidity of the monks as to what size a parish would be. We chopped the parishes around somewhat, but tried to keep them as the nucleus of each Rural Area. We found the GAA idea was a great help, because most of their teams were parish teams. . . their loyalty was owed to the parish and in this way we had rivalry between the parishes as to which should get the electricity first.'[1]

Owing to the time-span involved – it would take at least ten years to complete the scheme – it was perhaps inevitable that there would be this keen competition between parishes to get to the head of the queue. It was important that an easily understood and acceptable set of criteria to determine the order of development should be established. Otherwise there would undoubtedly be widespread pressure, political and otherwise, to have areas selected out of turn.

It was also essential from the point of view of the government and the ESB that with the large capital expenditure involved the work should be carried out in as orderly and economical a manner as possible. The government, in its approval of the scheme in August 1943, had stressed that priority should be given to the most remunerative areas with the stipulation that initially one area must be developed in each county. For reasons of efficiency and control, the work was decentralised within the framework of the ten ESB 'country' administrative Districts and thus, after the initial selection of one area in each county, the contest for priority was in fact between areas within each ESB District.

The term 'most remunerative areas' referred to the ratio between the capital cost of supplying the consumers and the annual fixed charge revenue yielded, which (with the help of the government subsidy) was intended to cover the annual fixed costs of the networks. When the selection process had been in operation for only a short time it was pointed out that proximity to existing electricity networks gave an area an unfair advantage even where the percentage acceptance in a more remote area was higher. The criteria were consequently modified so as to give the percentage of householders accepting supply equal weight with the percentage return on capital. It was therefore of the greatest importance in securing early selection that the maximum number of householders would agree to take supply. A large proportion of the farming community had yet to be persuaded that the advantages of electrification outweighed the costs involved. Many small farmers were still operating in a subsistence economy. The prospect of committing themselves to paying a regular fixed charge in summer as well as winter irrespective of their electricity consumption and of their financial situation was difficult to accept. In addition there was a widespread suspicion that electricity was dangerous.

On the other hand, in most areas a considerable number of householders were more than willing to sign up for electricity. As well as the more progressive farmers there were the clergy, teachers, shopkeepers, public employees and other rural dwellers who were more aware of the benefits of electrification, both to themselves and to their community as a whole. It was from these people that the initial pressure for electricity came and in most cases the early selection of an area depended on their success in persuading their more reluctant neighbours to sign up. The focus of activity in this field was the local Rural Electrification Committee.

Roe had quickly realised that his task would involve not alone motivation of the REO staff but of the rural community itself. If the scheme were to be launched without the involvement and commitment of the rural community it would fail. It would involve a change in ways and attitudes which would not be achieved by REO agents alone. A desire for change and improvement in the prevailing standard of living, and for availing of the coming of Rural Electrification to achieve this, must be fostered from within the community. Roe's involvement with Muintir na Tíre had made him sensitive to community reaction and thinking. Some lengthy discussions with Canon John Hayes (founder of Muintir na Tíre) convinced him that the way to motivate the consumers in the rural areas was by getting them to participate in the work. Thus was born the idea of the parish Rural Electrification Committee.

The make-up of these local committees varied. Sometimes they evolved from the local guild of Muintir na Tíre, sometimes from Macra na Feirme, the local co-operative society or the parish council; in many cases some local people simply came together and formed an ad hoc group. Every effort was made to avoid identifying the committee with any political party or denomination.

THE PRELIMINARY CANVASS

The main function of the committee was to carry out the preliminary assessment of demand for electricity in the area and submit this to the ESB in the form of a memorial signed by all householders prepared to install it. Where the response was highest an official canvass was then carried out by an REO Area Organiser who measured up the premises, determined the fixed charges and got the householders to sign official application forms. He was also in a position to explain the benefits and answer queries from householders some of whom had only the most rudimentary idea of electricity at that stage.

Naturally the local committee was motivated to obtain a high percentage of acceptances so as to ensure that its area qualified for early selection. There were two criticial stages in this regard. The first was when the original memorial was being prepared, as on this depended the promptitude with which the area was canvassed officially by REO. The response varied from 'negative' through 'doubtful' to 'enthusiastic'. Writing from west of the Shannon, Ora E. C. Kilroe reflects the typical reaction of the younger farm housewife:

> Two young men descend on us like a heavenly visitation. They are a deputation from the local Committee and want to know if we will take the electric! I can hardly believe my ears. What a boon and solace it would be and how miraculously it would lighten the unrelenting drudgery of a farmhouse. I walk about all day in a dream of immersion heaters, electric irons, churns, incubators and husband happily contemplates oatcrushers. We hear only one in the District has refused it, a hardy old dame of 85 who says indeed she will not have it and be setting fire to the thatch.[2]

A rather touching letter quoted in *REO News* in November 1951 is indicative of the latter approach, particularly among the older generation. It was written to a friend of the editor by an old retired servant of the family who then lived in a small cottage between Monasterevin and Rathangan:

Above: The ESB crew is welcomed and entertained by a grateful community. *Below*: Battery radio no longer required!

> ...the electric light is at last installed in far-off Clonmoyle. Kate H– could not stick the bright light at first and lit an oil lamp. I did laugh at her. They are all very grand until it comes to the paying for it and then they may get a surprise....

The old lady flatly declined the offer of her former employers to pay for the installation of light in her cottage and pay all the electricity accounts: she simply preferred not to have electricity.

A former Area Clerk recalled an elderly couple living on their farm in County Monaghan who would not take supply despite repeated visits from members of the local committee. Their only child, a daughter, had married, and her husband had moved in and was running the farm. However, the old man would not relinquish control up to the day he died. As expected, he left the place to his daughter but on condition that there would be a room provided for his widow. This request was of course honoured by the young couple who immediately looked for electricity supply. The most convenient point of connection was at the back of the house, outside the widow's room. She, however, would not allow electric wires to be connected to any wall forming part of her room. Connection was made to another part of the house wired for electricity — all except the widow's room. She continued to use an oil lamp up to the day of her death in fulfillment of what she interpreted as her late husband's wishes.

Not all the older people were so reluctant. In the Looscaun area near Woodford in County Galway in 1953 a seventy-eight year old bachelor, Dan Bonfield of Gurteeny, took supply into his tiny two-roomed house, possibly the smallest premises connected up to that time. His total installation was one light, so that he could play cards with his neighbours. He was very fond of a game and with the natural dimming of his eyes at that age he felt that an oil lamp or candle was insufficient if he was to keep a reasonable eye on his fellow-players.

THE OFFICIAL CANVASS

The second and more vital stage was when the REO canvass commenced and the householder was advised of the actual fixed charge on his or her premises and asked to sign an official application form. At this stage many who had signed the original memorial fell out, usually to the dismay and chagrin of the local committee, as a high percentage of refusals generally meant relegation to a low position in the league table.

The efforts of the committees and the methods employed to try to secure a high sign-up were varied. In many cases a further bout of intense canvassing was involved. Other methods adopted were more spectacular. The committee of the Knockbridge, Louth and Tallanstown areas of County Louth, for example, persuaded local businessmen to donate advertising space in the *Dundalk Democrat* for the purpose of exhorting the faint-hearted to sign up. The examples given show the vigorous style considered necessary to achieve results. (The second example shows the ability of the donor to interlace his own message with that of the committee.)

It is not possible to describe the workings of the local committees without referring to the role of the local clergy. For generations the local priest in rural Ireland was looked to by his congregation for leadership and advice. What was more natural than that the chairman of the local Electrification Committee should be the parish priest or curate (and for that matter that the secretary should be the local schoolteacher)? In general the clergy recognised the value of electrification and took a positive attitude in encouraging their parishioners to avail of the scheme. Indeed, in many cases the local PP personally took part in a round up of refusals among his flock when the canvass was not going too well. Phil Casley, one of the Area Organisers attached to the Galway District recalled that in the southern part of the county he was asked to accompany the parish priest on such a mission. It was obvious, however, during the exercise, that many of the householders who signed up did so because of their traditional reluctance to refuse a direct request from the PP. Nevertheless, one parishioner successfully avoided the issue by disappearing into the horse's stable with a large forkful of hay when he recognised the purpose of the approaching pair. By the time they had reached the door of the stable, all that was to be seen inside was a horse contentedly munching from the pile of hay. Of the wary householder there was no sign, nor did he subsequently reappear in any other part of the farmyard. The visit had to be written off as a failure. Subsequently the ESB man ascertained what he had at the time suspected, that a pile of hay can be excellent camouflage, given a co-operative horse.

Unfortunately many of the agreements obtained in this way did not last the course, especially in the early years. When the area was selected for development and the construction crews moved in, the moment of truth arrived for many who were faced with the choice of wiring up their houses and getting the supply connected or of backing down and opting out of the scheme. In some areas at this stage, as many as one-third backed down. The reasons given were many and varied, but for the vast majority it was a feeling that when it came to the crunch electricity was a luxury they could not afford. In the case of such people a former AO recalled:

RURAL ELECTRICITY - SIDE "LIGHTS" FROM HISTORY -

1 — 878 B.C.: The Firefly and Lantern-Beetle catchers of Abyssinia condemned the introduction of resin wood "splinters" for lighting. "Splinters" were **TOO DEAR & TOO DANGEROUS !**

2 — 134 B.C.: "Splinter" makers of Sweden condemned the introduction of rush "dips" - **TOO DEAR & TOO DANGEROUS !**

3 — 310 A.D.: The "dippers" - of Paris cried aloud against the use of tallow candles. These were - **TOO DEAR & TOO DANGEROUS !**

4 — 1660 A.D.: The Candlemakers' Guild of London protested against the introduction of the "convex oil lamp" - **TOO DEAR & TOO DANGEROUS !**

5 — 1819 A.D.: The oil lamp makers of Quito were "afire" against the introduction of coal gas - **TOO DEAR & TREACHEROUSLY DANGEROUS !**

6 — 1878 A.D.: The Coal Gas Company of Tibet (we're told) "fumed" against the introduction of electricity - **TOO DEAR & TOO DANGEROUS - ! !**

Householders of Tallanstown, Louth, Knockbridge, you are not living in Tibet and surely you do not want to be 70 years behind the times. Think then of something else to say against Electricity, for it is neither DEAR NOR DANGEROUS. This is our last earnest appeal and your last chance. Now or never. Sign your acceptance form without hesitation.

(This space has been generously donated to your Committee by Messrs. Macardle, Moore & Co., Dundalk Brewery)

KNOCKBRIDGE - LOUTH - TALLANSTOWN
RURAL ELECTRICITY

Deanna (Dunbin) writes : "I jilted the cad. He broke his promise to take Electricity"

More power to you Deanna. The mean rat was almost guilty of BREACH OF PROMISE. Don't worry dear : we have five of the handsomest - if shyest - young bachelors on our Committee. THEY'LL say it with ELECTRICITY if ...., and will get the complete outfit for the day of "daze" from PETER B., who has so generously donated this space to their COMMITTEE.

HERE I AM, YOUR PETER "B"
I'M KNOCKED OUT BY THE E.S.B.
BUT AS WE BOTH GIVE YOU DE-LIGHT
PLEASE SUPPORT US DAY AND NIGHT

YOU WILL KNOW
WATT'S WATT

If you buy your clothes at
26, Clanbrassil St., Dundalk.

Advertisements for the rural scheme published in the *Dundalk Democrat*.

> Some organisers felt misgivings, especially in a house with old folk, to see them torn between a desire 'to help the priest' and their genuine fear of tying a new 'Ground Rent' around their necks. However, these misgivings quickly dissolved when one called on the same people after they had been 'switched-in'. Their obvious delight with electricity had completely overcome their fears. It was amusing to hear them comment about cobwebs which they had never known existed.[3]

A secondary reason in the case of the older people was that it was somehow dangerous. The *Evening Herald* (7 February 1953) carried a story from the Annies district of Scotshouse near Clones where an old man paid his customary weekly visit to his sister who had just been connected up. He watched with fascination as she operated the new electric iron and boiled water in the new electric kettle, but stubbornly refused to drink the tea made from the latter, as he believed that the water was electrified. Willy nilly, she had to make fresh tea from water boiled in a traditional kettle on the turf fire.

As times improved, as the value of electricity to their neighbours became manifest and as their original fears receded all of these householders almost without exception eventually applied for connection.

Not all the clergy were in favour of the scheme. Here and there one found a parish priest who had fixed ideas about how his parishioners should live – as their forebears had done – and who suspected that electricity was only the first step in a radical and possibly unwelcome change in their lifestyle. Others considered that it involved the people in unwarranted expense. In a parish not very far from Cork City, the parish priest was opposed for this last reason. If a large number of the parishoners found it difficult to pay their church dues, as appeared to be the case, how could they afford this new expense? On the other hand, his curate, a young man, was so committed that he was accused by his superior of preaching rural electrification on Sundays instead of the Gospel. However, the work went ahead and towards its completion the parishoners were anxious to have a formal 'switch on' ceremony in the local hall. It was decided for obvious reasons to hold this while the old man was away on holiday. In the event, he returned unexpectedly as proceedings were getting under way, no doubt having been alerted to the goings-on by a loyal parishoner. He strode on to the platform where the various members of the committee were assembled. Facing the audience he declaimed 'Ye have it now and I can tell ye it's going to be dear, and mighty dear, but ye would have it and ye can now go and pay for it!'. Even in this case, not many months had passed before the reverend gentleman himself had been converted to electricity. In another parish the PP was adamant that he would not take the supply until the canvasser slyly let drop the information that the bishop was very keen on the scheme. Standing up and striding to the table he thumped it, exclaiming 'Well, what's good enough for Jim R– is good enough for me. I'll take it.'

In general, however, clergymen were very aware of the opportunity offered by electricity and the necessity to grasp it. Many worked very hard with the local committee and with the REO crew when it arrived, to ensure that in their locality as many householders as possible would take supply and avail of it to improve their living standards. In the final report on Ballivor Area in County Meath the RAE refers to the help given by Fr Kiernan, the parish priest:

> From the first day of arrival he gave the construction crew every assistance, got accommodation for them and on the spiritual side with the help of the Area Organiser Barney McEneaney arranged a special retreat at times most suitable for their working hours. If we ever decide to have a patron priest of rural electrification he is readymade.

One of the veterans of those early days, Noel McCabe, in recalling the event to the writer added:

> To ensure the maximum attendance Fr Kiernan visited the local hostelries and 'requested' the owners to close down for an hour each evening while the retreat was in progress. However, being a just man he also 'requested' the local Sergeant not to visit the pubs for a corresponding period after closing time!

With the arrival of the area crew in the parish the prospect of electrification at last became a reality and a new flurry of activity began. Suitable premises had to be secured for the area office, as also had a stores and outdoor storage compound. Digs had to be provided for the travelling members of the crew. House wiring contractors had to be located and householders encouraged to commence wiring their premises. This last was the almost irrevocable commitment and many of the faint-hearted fell out at this stage.

The Area Organiser and the field design engineer usually arrived some weeks ahead of the rest of the crew to confirm the final 'acceptance' position and ensure that the pegging out of the lines got well ahead of the construction gangs. Wayleave notices had to be served on landowners to cover every pole and house service, and any disputes arising had to be cleared up before work could begin. This latter task was usually carried out by the Area Organiser who had to call on every householder in the area. Sometimes there were problems of identification, as in the Clonark area in County Roscommon, when the Area Organiser found one group of eight householders all of whom were named McManus. A similar problem arose near Glengariff in the Castletownbere area where eight householders connected to the same transformer in the townland of Firkdale were all found to be O'Sullivans. For the record, the eight first names were Johanna, Humphrey, Edward, Mrs W., Patrick J., Patrick E., Michael J. and Michael P. Óg.

This first visit of the organiser had many objectives. He ascertained if the householder was accepting supply and, if the answer was negative, usually employed his persuasive powers to try to effect a change of mind; he urged him or her to get the house wiring under way and advised on provision and position of sockets; he served wayleave notices to cover any poles on the householder's land and negotiated changes in proposed pole positions where necessary to avoid obstruction to farming activities. At a later date he would return to check on the progress of the wiring and to demonstrate and possibly sell some items of electrical equipment, but this first programme of visits gave him a valuable overview of the total scene and of what problems were likely to arise during the construction period.

There was frequently a long period of time between the original canvass and the eventual arrival of the construction crew. In many cases the Rural Electrification Committee, which had been so active in those early days, had been dissolved, its main object now achieved. In other cases, however, the members were still active and concerned to ensure that all went well. These afforded tremendous help to the crew in using their influence to prevent householders from backsliding, in locating lodgings and office accommodation,[4] recruiting local labour, organising the house wiring and resolving many of the problems that inevitably arose on a project having had such impact on people's property and way of life. Even when the formal committee no longer existed, an informal group, including the local clergy, schoolteachers and prominent local citizens, frequently emerged and provided very valuable support and liaison with the community in general.

The newcomers were welcomed. Even in areas where the lodging of visitors was a rarity there was seldom any difficulty in obtaining accommodation for the ten to twenty people who formed the skilled nucleus of the team. For many crew members of urban background the experience of rural life and rural hospitality thus gained provided a new and enriching dimension in their lives and the years spent on 'rural' were looked back on with considerable nostalgia.

This mixing of the visiting members of the crew with the local community and the recruitment of local men for the duration of the job did much to ensure the acceptance of the scheme as an integral part of community development, and many small farmers had their meagre incomes supplemented by earnings on the scheme. Indeed for many farmers' sons it was their first opportunity to earn money in their own right.[5] The result was an unparalleled degree of goodwill and co-operation. Despite the necessary encroachment on the land and property of almost every householder, disputes were a rarity and where they arose were usually solved on the spot.

The ESB staff on their part responded to the friendly welcome by entering into the local scene with enthusiasm. Jim Wolahan, a former Area Clerk recalled his stay with a farm family in their house in County Monaghan and how, on returning to his digs at night, when the family had gone early to bed, he would find his supper laid out on the kitchen table. His first duty, however, was to take a feeding bottle of milk to a piglet – the delicate runt of the litter – which was cosily ensconced in a canvass bag hammock beside the fire. Only when the bonham had been fed and tucked in for the night did he commence his own supper. In being entrusted with this important chore he felt he had been completely accepted as one of the family.

The people of Blackwater, Co. Wexford, must have felt that the Area Organiser Phil Cox was one of their own as they saw him leading the parish priest's donkey and cart down the village street laden with a table and press to equip the newly opened area office, which their energetic committee had miraculously located. The same man became even more integrated into the community when he married a local girl. The newly arrived RAE in Ballivor, Co. Meath, Angus Ryan, looked down the village street and anxiously enquired of the local Area Organiser 'Who is the mot?' Some months later he led her to the altar. Over the course of the scheme, the number of ESB staff who first met their future wives while on rural electrification work ran into very large figures. The *Kilkenny Journal* of 14 September 1950 referring to the departure of the crew from the completed Gowran area said 'Gowran will miss them; these boys were drafted in here total strangers less than six months ago and promptly won the friendship and confidence of all and the hearts of some'.

In the evening the pubs and dance-halls were enlivened by the newcomers who brought to the local scene their own particular style. Some had earned their place in the travelling crew by virtue of their fabulous strength or physique – they could fling poles around like matchsticks – and delighted in demonstrating their porter-drinking capacity. Others were more interested in the local women. Many had musical or dramatic talent and proved a great asset to the local societies. The advent of electricity to the parish hall was frequently the occasion for a celebratory concert or play. One crew operating in the west Cork area staged many such concerts and plays from its own resources in aid of local worthy causes.

While the visiting crews were generally well behaved, there was the odd brawl. Often this was over local girls, who to quote an old 'rural' hand 'were always glad to welcome new blood'. It had, he explained, been the same with the army during the war and the visiting rural electrification crew were regarded as worthy successors. The local boys employed for the 'duration' also made their presence felt in the village on pay night and inevitably there were clashes. An RAE in County Sligo was horrified on opening a copy of the *Champion* to see the headlines of a case in the local court: 'ESB Men Convicted of Assault'. With visions of his visiting team being run out of the district he made enquiries to find that the men in question were natives of the locality employed as short-term casual labourers on the scheme. When asked for their occupation for the purposes of the charge, they had obviously felt that 'ESB men' gave them a cachet superior to that of 'farmer's son'.

As work in the area came to an end there was a phasing out of the local temporary staff. There was a certain loss of face in being among the first to be let go and reaction was on occasion dramatic. In an area in County Tipperary one such aggrieved worker expressed his displeasure by coming in to the village in the evening with a shotgun, which he discharged at intervals while loudly calling for the supervisor Martin Whyte to show himself. The latter recalled the incident thirty-five years later: 'I had been feeling tired after a hard day but when I heard the shots the tiredness dropped off me and I ran in fear of my life'.

Another turbulent stalwart – a member of the tree-cutting gang – raised his axe and, to the accompaniment of some lurid language, buried it almost to the hilt in a convenient tree-trunk when told he had got the 'chop'. The supervisor had taken the precaution of waiting until he was on the opposite bank of a river before imparting the unwelcome news.

Such incidents were, however, the exceptions. Usually the parting was made with good humour and expressions of goodwill on both sides. In some cases one or two of the most promising local lads were invited to travel with the crew to the next area as 'semi-skilled' workers, which was often the first step on the ladder of promotion.

The easy integration of the visiting crew members with the local community did much to soften the traditional reluctance to change and the misgivings that might otherwise have existed. In the Lispole area in County Kerry the roving reporter of *The Kerryman* wrote of casually dropping in on the 'switch-on' and referred to the high opinion the local people had of the construction crew, particularly of the business-like manner in which they tackled their job and the excellent friendly relations that existed between them and the residents of the locality.

Perhaps the tribute paid by the parish priest in Carnaross, County Meath, said it all. At Mass the Sunday before the last of the crew departed he publicly wished them farewell: 'I was very moved to see the way in which these nice people came amongst us, did their job with speed and efficiency that opened our eyes after our experiences with other official bodies, behaved quietly and decently and left without any fuss or display'.

Canon John Hayes,
founder of Muintir na Tíre.

CHAPTER ELEVEN

Problems

BACKSLIDERS

IF ONE WERE TO ASK a Rural Area Engineer or a Rural Area Organiser who worked during the first decade of the Rural Electrification Scheme what caused him the greatest problems, the answer would probably be backsliders, those householders who had signed application forms during the official canvass of the area but who, when the construction crew arrived, refused connection or had still not been connected when the area was completed. The design and costing of the electricity network for each area were carried out on the basis of the signed application forms. The estimated return on capital for the area, which, together with the percentage of applications determined the position of the area in the selection table, depended on the estimated fixed charge revenue from the acceptances. There were thus many undesirable consequences to backsliding. The withdrawal of a householder at this late stage generally meant loss of expected revenue without a corresponding saving in capital expenditure. The loss of even one household in a group frequently resulted in others moving into the 'uneconomic' category and so further reduced the number of houses connected. If, as frequently happened, backsliders appeared in large numbers, the economic return from the area could fall drastically. When this occurred, householders in areas still awaiting selection and perhaps only slightly lower in the table could feel aggrieved and make accusations of unfair treatment. An appreciable number of areas with heavy backsliding could significantly alter the overall economics of the scheme at any particular time and call into question many of the assumptions on which it was based. Backsliding was consequently a very serious problem, which had to be tackled and solved unless the future of rural electrification was to be put in jeopardy.

To the young Rural Area Engineer, this situation was as unexpected as it was difficult of solution. While courses in the engineering schools had no doubt covered adequately the technological problems of electricity distribution, associated problems based on human nature or human reactions to innovation were unlikely to have been envisaged, still less treated. A totally new dimension was thus added to his job — the development and exercise of his human relations skills. This applied even more so to the Area Organiser who was in the front line in this battle, calling, as his job required, on every householder in the area and arguing, persuading, cajoling him or, less often, her into reconsidering their withdrawal.

It was first necessary to pinpoint the cause or causes of the backsliding. Going back on one's word had never been a characteristic of the Irish rural dweller: the commerce of fairs and markets depended on the tradition that a man's word was his bond. Yet the phenomenon of backsliding was widespread in the first decade or so of the scheme. The situation improved with time but in the late forties and early fifties many Rural Area Engineers were aghast when the time came to pack up and move on, to find that they had connected only a fraction of the number of consumers expected and budgeted for.

In Easky, Co. Sligo, for example, at the end of the construction period in May 1948 only 210 out of an expected 518 had been connected (although by the following December this had risen to 345, which was still only two-thirds of the original number who had signed up). In Tara Rural Area, Co. Meath, in 1950 the final report of the RAE recorded seventy-seven backsliders out of an original acceptance of 451 householders. In this case twenty new consumers took supply. The report for Dromohane, Co. Cork, recorded fifty-seven backsliders, mostly cottages and small farms, out of an original figure of 320. It also notes that twenty householders who had not signed up originally had now joined the scheme and that these were the larger type of farmer. The RAE in Grange, Co. Sligo, noted seventy-three backsliders out of an original figure of 460 and (here is a clue) identified them as mostly old people or small farmers who had originally signed 'in a burst of enthusiasm'. In this area, however, the score was balanced somewhat by fifty-six extra consumers.

On the other hand, considerable gains in the number of consumers connected were achieved in some areas. In the Dundrum area in County Tipperary, while twenty-six householders out of the original 340 backslid, sixty-five new consumers were obtained. The closing report refers to one backslider who 'had built his tiny bungalow on a raft which he rolled away out of the area leaving behind him only a number on the map, a signed application form and a most indignant Area Organiser'. In many areas the number of 'new' consumers connected up outweighed the backsliders, although in most of these cases an intensive redesign of the

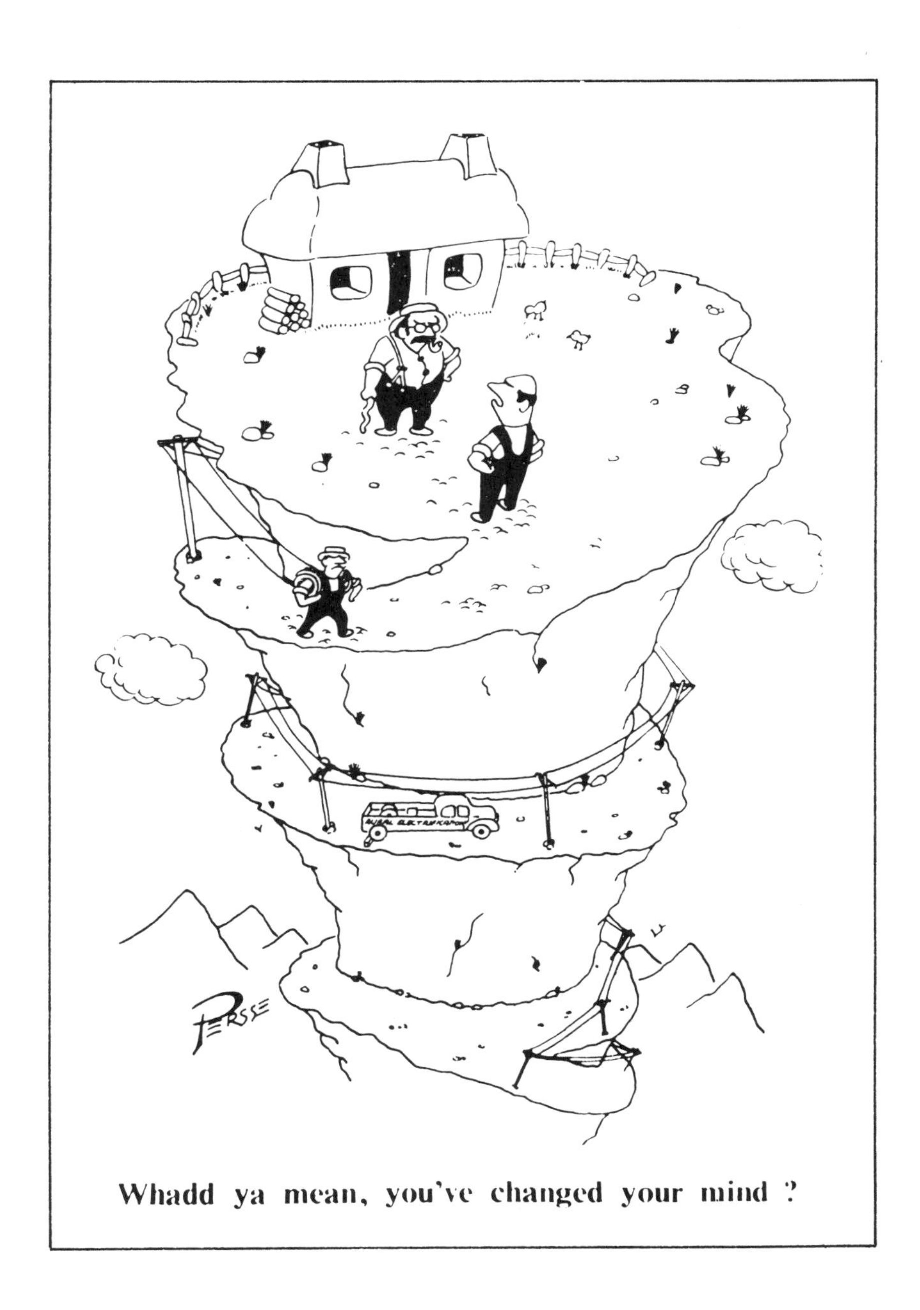

Whadd ya mean, you've changed your mind ?

networks was required to minimise the extra capital required. Even when there was heavy backsliding it was frequently expected that this would only be temporary. The report from Nobber, Co. Meath, where at the completion of construction work there were ninety backsliders, observes that 'in many of the cases, only the head of the house had a serious objection and was being subjected to considerable pressure from other members of the family'.

In areas adjoining large towns and cities it was to be expected that backsliding would be minimal and this was the case in Monkstown in Cork where the consumers connected increased from 502 to 676, giving a return on capital of 11.4% ('the highest in the country'). In Cloughran adjoining Dublin Airport the figure went from 165 to 294. However, equally dramatic increases were found in unexpected locations. In the Achill area, while there were seventy-seven 'completely recalcitrant' backsliders, which caused three others to become uneconomic, ten houses that had become vacant and one which had been demolished by lightning, 244 new consumers were obtained to give a final figure of 804, 'due', in the words of the final report, 'to the unrelenting efforts of the Area Organiser, Phil McGovern, who maintained his efforts where men of lesser courage and industry would have admitted defeat'. In Ballinacargy, Co. Westmeath there were ten backsliders out of 182 original acceptances but fifty-four additional acceptances were secured. The backslider figure 'would have been much higher but for the efforts of Rev Fr Kearney, CC, in bringing in the most recalcitrant [that word again!] of the refusals'. He was tireless in his efforts and his assistance is summarised in the final report, 'I have no hesitation in acclaiming him as the best AO of all time'.

West Donegal, one of the least prosperous parts of the country in 1952 might be forgiven if the number of backsliders was higher than normal. However, to quote from the closing report for the Kilcar area: 'The original total acceptances was 223 and the final total, 323... . There was only one backslider and even this one had been connected up since the job was completed. An amazing characteristic of the area was the enthusiastic way in which the householders took to the idea of rural electrification although the vast majority of them had little idea of it beforehand. This is particularly surprising when we learn that farming as such is almost non-existent in the locality, there being an overall average of approximately 6 acres of arable land per holding... .' This contrasts with the situation in the opposite side of the county: in the comparatively prosperous area of Manorcunningham there was a net loss of 137 consumers, with a reduction in fixed charge revenue of over £700 per annum. The figure included 132 backsliders, rendering a further thirty-three houses 'uneconomic'; fourteen houses had become vacant since the canvass and forty-two new consumers were obtained.

One simple reason for backsliding stands out: an over-enthusiastic canvass by the local committee. The greatest proportion by far of backsliders came from the categories of small farmers and agricultural labourer. These households very often operated on a semi-subsistence economy where regular payments of cash for services were not a tradition and where income was either very low, very irregular, or both. Many then regarded electricity as an unwarranted luxury and recoiled from the concept of paying out a regular fixed sum irrespective of the use they made of the service, which in many cases would be confined to winter illumination. The problem facing the local committees was that in many areas these reluctant householders comprised a high proportion of the total and unless they were persuaded to sign the relevant application forms the chances of early selection of the area for electrification were poor. It is little wonder then that pressures of all sorts were exerted to bring the acceptance figure up and in very many cases it was a case of signing 'to bring the electricity to the parish'. When, with the arrival of the construction crew, the moment of truth also arrived, necessitating an investment in house wiring, many of the reluctant signatories decided to opt out. Much time and effort by ESB staff and by local helpers were put into trying to recover the situation.

Not all backsliders fell into the above category. There were many reasons for refusal or delay in accepting supply; in the case of public buildings it might be a simple case of bureaucratic delay in wiring the building. In this respect the *Kilkenny Journal* of 14 September 1950 takes the Government itself to task. The Gowran correspondent writes:

> Those in high places have not spared their energies in exhorting the country people to avail themselves of the wonderful benefits of Rural Electrification and urging them to make no delay in installing electricity, yet the only two institutions in Gowran which come directly under the Government – the Railway Station and the Garda Barracks – are left without the 'wonderful benefits'.
>
> I don't know what conditions prevail in either system but certainly the example is not very inspiring and speaks poorly for the sincerity of the Government push behind Rural Electrification.

In the case of the 132 backsliders in Manorcunningham an analysis was given by the RAE. Included were twenty-seven farms, seventy-nine tenants' houses, eighteen small owner-dwellings, two churches, one school, three shops, and one hall. Of the twenty-seven farmers, two refused supply, 'so as to avoid poles on their land', but the remainder gave no decisive reason. Of the seventy-nine tenants, thirty-seven refused because the landlords would not wire the houses and the remainder were occupied by labourers who felt they could not meet the expense. This latter reaction was echoed in 1953 in the Kildangan area in County Kildare, which included in its total of 509 premises 196 labourers cottages and a further one hundred or more houses of similar or smaller size. In spite of the fact that Kildare County Council wired the cottages, to quote the closing report, 'most of the tenants were not at all enthusiastic about supply and the construction crew were faced with the appalling total of 106 backsliders'. The report goes on: 'It was quite clear that the area has been subjected to a considerable amount of local pressure prior to and during the canvass and as many of the residents are only in part-time or seasonal employment it is not surprising that the results were as bad as they actually proved to be'.

In Rathoe Area, Co. Carlow, in 1951, where there were sixty-eight backsliders against a gain of sixteen extra consumers, a detailed analysis of the former was carried out by the RAE. The results were interesting:

No reason given	22
Occupier in poor financial state	19
Houses had fallen into bad repair	8
Houses had become vacant since canvass	6
Occupier had died since canvass	4
Board of Works refused to wire Garda barracks	2
'Economic' consumers had become 'uneconomic' due to 'backsliding'	7
Total:	68

Another survey of the unconnected householders in an area in the midlands that had been developed for some time gave some interesting results:

Waiting to see how charge would work out with neighbours	70%
Rendered uneconomic due to neighbours refusing connection	20%
Originally refusing because of illness, etc.	10%

Action taken to try to reduce or eliminate backsliding included personal visits, usually by the Area Organiser, mostly alone but sometimes accompanied by members of the Committee, public meetings, demonstrations of equipment, threats to withdraw from the area or to take down parts of the network. Intensive canvassing of householders who up to then had refused to join the scheme often had the result of getting new consumers to offset the losses. In Bekan, Co. Mayo, an area where the average farm holding was fifteen acres, it appeared at first as if upwards of a hundred householders were going to drop out. A series of personal visits by the Area Organiser, however, reduced the figure to twenty, and in addition 138 extra consumers were gained.

Sometimes the local clergy helped. The RAE in his final report on the Kildalkey area in County Meath speaks in glowing terms of Fr Kiernan, the helpful parish priest of Ballivor who also had care of Kildalkey, and attributes a gain of sixty-two new acceptances to his efforts. In the Castlegregory area in west Kerry, on the other hand, an appeal was made to the local clergy, but, to quote the final Area Report, 'though they exhorted their people to accept supply, they themselves became backsliders'. 'Apart from this,' the report goes on to say, 'the position was not too bad'(!).

The extra consumers obtained as a result of hard work by REO staff during the construction period did much to offset the backsliding. An analysis made of the position at 31 March 1952, when 180 areas had been completed, showed that the numbers actually connected slightly exceeded the original number of 'economic acceptances' for these areas — 56,818 connected as against 56,332 original economic acceptances. This represented, however, only 60% of all premises (93,970) in these areas and indicates the large number of unconnected premises still remaining which would have to be tackled under the various 'Post Development' campaigns.

THE 'GROUND RENT' CONTROVERSY

The decision in the report of 1944 to recommend a two-part electricity tariff for rural electrification was not taken without full consideration of all other options. The main factor in favour of this approach is that the annual cost of providing electricity can be divided logically into two parts. One is the annual cost of the distribution network itself which is made up of the annual capital charges, operation and maintenance costs, meter reading and billing costs and administrative costs. Once the network has been built and consumers supplied these costs are incurred irrespective of the amount of electricity used by the consumers. They are 'fixed' to this extent, but of course must in time respond to changes in the costs of the various components. The second part is the generation and bulk transmission costs which are best expressed in the form of costs per kWh or per 'unit' of electricity, as they depend in the main on the actual amount of electricity generated and transmitted in bulk to the main feeding points.

For a developed and stable pattern of electricity usage the total costs may, without undue difficulty, be charged for solely on a unit basis without a fixed charge component. For a scheme such as the Irish rural electrification scheme where the pattern of use would have to be built up slowly over the years, such an approach could lead to financial disaster by actually inhibiting the development of the use of electricity from the start. In this regard the 1944 report quoted the experience of Sweden, one of the pioneers in rural electrification in the twenties. Much of the distribution was left to autonomous local rural distribution societies which in an effort to meet the immediate expressed desires of their members failed to make proper provision for the long term.

It was the study of such cases which convinced the ESB of the necessity of using a two-part electricity tariff system. Without this system it is doubtful if the rural electrification scheme would ever have got off the ground. The fixed portion of the charge based on the floor area of the house and out-offices quickly became known as the 'ground rent', and among rural dwellers and rural politicians was for many years to be the most hotly debated aspect of the scheme. To consumers who had been brought up to pay for their light and heat more or less as they used it, it appeared unreasonable to require them to commit themselves to paying a substantial annual sum merely for the *provision* of supply. They had no objection to paying the unit charge for the electricity actually consumed and indeed would not have objected, in the short run at least, to a higher unit rate if this would eliminate the fixed charge, feeling no doubt that this would allow them more control of their expenditure. The fact that the annual fixed charge (usually from £6 to £9) was divided into six equal two-monthly instalments made things worse. Rural consumers foresaw themselves paying the fixed charge in summer, when they were using little if any electricity.

The farming community had only recently won its freedom from the clutches of rack-renting landlords. Now its members saw themselves faced for the indefinite future with a new series of regular and substantial payments to the ESB based on the size of their houses and outbuildings and irrespective of how much electricity they used. There was only one term they could apply, 'ground rent'. It made no difference how often or how emphatically the ESB officials explained that it was a charge to meet the annual network costs, and that these costs did not depend on the amount of current used and were best met in this manner. Having so met these costs the actual costs of current itself could be kept low. It could not therefore by any criterion be called a ground rent. It was all to no avail. The term stuck, particularly in the west.

At times, indeed, this 'ground rent' was directly identified with the land rent paid to the old landlords and proposals to increase it brought similar reaction. Towards the end of 1960 the ESB, against a background of rapidly rising costs introduced a 10% increase in fixed charges, the second 10% increase since 1946. The reaction country-wide was vigorous, with protest after protest being made. That it was a strongly emotional issue was shown by the report in the *Western People* of 11 March 1961. A mass meeting of five hundred farmers in Crossmolina passed a resolution, which was sent to the Chairman of the ESB by the Honorary Secretary of the National Farmers' Association, North Mayo Regional Executive:

> The farmers of North Mayo protest against the raising of the ground rent by the ESB and the increased charge for electricity. The increase in the ground rent has brought the small farmer from Donegal to Kerry back to the days of the Clanricardes and the Boycotts. It is unchristian that the lowest wage earners in Europe (see Farm Survey 1957) should have to pay for the most expensive electricity west of the Iron Curtain. The £3,000 a year man in the city with a seven-roomed house has a cheaper ground rent and cheaper electricity than the cottage in Erris or Connemara. To rob the poor to pay the rich seems to be the new policy of your Board. It is not surprising that 252 of our homes have disappeared from the Western Seaboard. What the crowbar brigade failed to do the ESB is helping to accomplish. We, the farmers of North Mayo pledge ourselves to use every means in our power to fight this injustice.

A similar resolution was forwarded to the Taoiseach.

In view of the fact that one of the primary objects of the Rural Electrification Scheme was to prevent rural homes disappearing and that Irish electricity was by no means the most expensive, the Chairman considered that the meeting had over-reacted somewhat to the situation. In a very detailed reply, made the following points.

The scheme was heavily subsidised by both the taxpayer and the [urban] electricity consumer.

Escalating costs had made increases in charges unavoidable. These had been kept as low as possible. The average fixed charge increase was only 3*d*. per week.

A reduction from eighty to sixty units, which had been simultaneously introduced in the size of the initial high-priced blocks of units, went far to mitigate the increase in the fixed charge. For consumers using, say, between ten and twenty units a week the average overall increase would be less than 1*d.* a week, and even for consumers using fifty units a week the increase would only amount to about 2½*d.* on average.

These increases scarcely warranted comparing the ESB with Clanricarde or Captain Boycott, but the reactions of these western farmers did indicate their emotional identification of the ESB 'ground rent' with the exploitation of their forebears by landlords and their agents. Their determination to resist collectively echoed the heady days of the Land League and the Plan of Campaign.

Even the efforts of the ESB to emphasise the term 'fixed charge' instead of 'ground rent' backfired: if it was a 'fixed' charge it could never be altered. A letter from the Secretary of Muintir na Tíre Branch, Glinsk, Ballymoe, Co. Galway, dated 11 January 1957 (following the first 10% increase in the fixed charge) reads:

> On behalf of the Glinsk Guild of M. na Tíre, I am requested to write to you as regards the increase in the ground rent. In this area we were led to believe that the units may be raised but that the ground rent would never be increased. Therefore if you insist on the increase, we request that the whole area be switched off.

There was a rather more lengthy protest by another gentleman, the central paragraph of which deserves quotation:

> Now Sir, the users that you have got into your net or the Farming Geese, as I might call them, have a few old manure bags to hold the feathers and I'm blowed if they will let you take the goose. They are looking for a reduction in the G R charges and unless they get it they will get out the meters. The ESB will get no revenue and the poles and wire may remain a memorial to the ESB. It would not pay them to take them down. Think it over — a half loaf is better than no bread. My rates have gone up £10 within the last 15 years. Add £8.9*s.* ESB charges and the loaf 1*s.*1*d.*, butter 4*s.*4*d.* You can call it government by the people or whatever you like, but the people won't stand for it much longer. They have borne the thing too long... .

Similar protests arrived from ICA guilds and NFA branches, all to the effect that it was understood that the fixed charge meant just that and would never be increased. Some claimed that a guarantee was given. It is worth pointing out again in these days of galloping inflation, that the amount in question averaged about 3*d.* or slightly over 1p per house per week.

There was another problem regarding the basing of the fixed charge on the floor area of the dwelling house and, at a lower rate, of the out-offices. Where estates had been broken up by the Land Commission, one of the incoming owners usually got the original large house and out-offices with his land allocation, while the other owners had new, smaller houses and out-offices built, more matched to the size of their land divisions. While the farm size and general income and electricity requirements might be the same, the occupant of the large house had to pay a higher fixed charge than his neighbour. A typical case from County Meath was raised by the NFA in 1958. One farmer had fifty-eight acres, as part of a five hundred acre farm which had been divided. He had got the original house and out-offices for which his two-monthly ESB fixed charge was £2.12*s*.6*d.* One of his neighbours had 150 acres, but because of a smaller house his fixed charge was only £1.7*s*.3*d.* Another neighbour with fifty-eight acres was paying only £1.3*s*.0*d.*

Another anomaly created by fixing the charge on the floor area was in the case of Protestant clergymen in rural parishes. They normally had to live with their families in huge rectories and manses, inherited from more spacious days. Their salaries were extremely low and their electricity demand was usually quite modest. Nevertheless, they had very high fixed charges, much higher than most of their neighbours.

Unfortunately it was not possible to devise a formula which would resolve such anomalies without creating even more. However, as inflation took hold and the cost of the unit escalated, and as electricity consumption increased, the contribution made by the fixed charge to the total electricity bill steadily got smaller and eventually ceased to be a major bone of contention.

CHAPTER TWELVE

Switching On

IN EVERY AREA, the high point of the scheme was the 'switch-on' of the first consumers. Later, when up to one hundred areas per year were being connected, this had become almost a routine and merited only a few paragraphs in the local paper. In the early years, however, the connection of a remote parish captured the imagination of the media and very full coverage was given, not only in local but in the national newspapers and indeed on the radio. REO, appreciating the value of such publicity (particularly in the persuasion of 'doubtfuls' and 'refusals' to take supply) co-operated enthusiastically with the local people in mounting a ceremony which would do justice to this important milestone in the life of the parish. Prior negotiations with the local authority had usually resulted in street lights being approved and erected. Traders had installed shop-window lighting and all the houses in the village had their internal wiring completed, with all bulbs installed ready for the great moment.

The typical scenario was a gathering of all the inhabitants in the village hall. On the stage was mounted a large switch, around which gathered leaders of the community: the local clergy, TDs, county councillors and one or two senior representatives of the ESB. If the occasion was considered to be of sufficient importance, a government minister was present. It was usually arranged that the switch on the stage simultaneously switched on all the village houses and street lights, as well as the lights in the hall. Where this was not feasible, a signal to strategically placed operators achieved the same result. In the hall itself a number of portable high-powered lighting battens was usually installed for the occasion to reinforce the conventional installation. Speeches were made, paying tribute to the organising committee and to the ESB staff. The historic importance of the occasion was noted and hopes were expressed that the coming of the light was symbolic of the dawn of a new era of enlightenment and prosperity for the community. Almost invariably the opening verses of the Gospel of St John were quoted '... and the light shineth in the darkness and the darkness hath not overcome it…'. A blessing was given, frequently, especially in the northern counties, jointly by the clergy of the different denominations, and the guest of honour was then invited to press the switch. One could anticipate a gasp of delight and a thunderous applause from the assembled crowd as the brilliant lighting took over from the existing candles and paraffin lamps. The gloomy hall was lighted as if by the noonday sun. Outside, the new public lights lit up the village street, while the brightly-lit shop windows added to the general gaiety. Far out into the countryside, bright pinpoints of light commenced to twinkle like stars against the blackness of the night as the rural 'spurs' were switched on. Heroes of the hour were the members of the ESB work crews, many of them local lads who had put in long hours and extra effort to ensure that supply was on time. Frequently the local committee invited the whole crew as guests to a meal and a dance.

Each 'switch-on' ceremony had its own particular flavour. The *Dungarvan Leader* of 28 August 1948 reports on the switching on of Ballyduff:

> Ballyduff was 'lit-up' in every sense on the night of Friday 20 August when the great 'switch-on' ceremony of Rural Electrification was performed. The ceremony was timed to start at 9.30pm and for some hours before that it was obvious that the people of Ballyduff and surrounding districts intended to make it a great night and one to be remembered. People arrived on foot and by car from all parts of the district and from Lismore, Fermoy, Tallow, Conra and Araglin. The greatest possible interest was taken in this, the culmination of many months of organisation by the Ballyduff Guild of Muintir na Tíre and hard work by the staff of the ESB. One old resident stated that he never saw a night of more spontaneous enjoyment. Everyone was happy and realised what a great night it was for the district when light and power, which is going to revolutionise country life, was made available.

The switch-on in Bansha was reported in the pages of *REO News* in June 1948:

> Bansha 24 May: Rev Fr J. M. Hayes, P.P. Bansha, switched in the village of Bansha from a platform on the Main Street. In spite of heavy rain, there was a large attendance of people from the two parishes of Bansha and Knockmoyler, accompanied by the Bansha Fife and Drum Band. Afterwards the local Guild of Muintir na Tíre entertained the REO construction crew to supper in the schoolhouse. Members of the crew contributed to the musical programme which followed. Speaking at the ceremony, Fr Hayes emphasised that '… rural electrification … is more

than an amenity — it is a revolution which will sweep away inferiority complexes'.

Again, the same issue of *REO News* reported the switching on of Abbeyshrule:

> Abbeyshrule 27 May: As no suitable hall was available, the local committee provided a large marquee for the occasion. Over 1,000 people were present to see the Minister for Justice, General Sean MacEoin, switch on. A demonstration [of electrical appliances and their use] followed to what has been our largest audience to date. The subsequent supper was prepared by members of the Irish Countrywomen's Association in a temporary electric kitchen erected in our demonstration van.

The *Longford News* of 29 May 1948 reported the proceedings at length, especially the speech of General MacEoin:

> It is now 25 years since the idea of the Shannon Scheme was first mooted by a youngster who is today a Minister — Mr McGilligan. There were misgivings by well-intentioned people at the time and it was even called a 'White Elephant'. To-day, we know the difference. We could do with five or six such white elephants. Abbeyshrule to-day is the envy of all Ireland... .' Preparing to switch on for the first time, the Minister said amidst a hushed silence 'may God give the Light of Heaven as I am about to give you light now'.
>
> All around became instantly illuminated and the applause lasted several minutes.
>
> An ESB official then demonstrated a number of machines... . His talk and illustrations were followed with keen interest by a crowd of several hundreds — there were a few idiotic interruptions by a local wit – or half wit – and an occasional echo of 'the loud laugh that ...'. It was a pity that the demonstrator hadn't got an electric chair handy!

In a lengthy article headed 'Kilmessan's Night of Jubilation', the *Meath Chronicle* of 10 December 1949 gave full coverage to four separate functions held on the same evening to mark the coming of electricity to the locality. The first was in the Temperance Hall where Liam Cosgrave, Parliamentary Secretary to the Taoiseach officially switched on the supply:

> The hall was lighted by candles, lanterns and paraffin lamps and truth to tell it was far from ineffective. When Mr Cosgrave touched the switch which illuminated the hall with the new current, there was an outburst of applause which could certainly have been heard on Tara.

Above: The 'switch-on' — the large metal-clad switch used for such events was 'liberally bespattered with rust spots from frequent former applications of holy water'. *Below*: Switching on the 100,000th consumer, March 1954 — the Ballinamult Creamery, Co. Waterford, floodlit for the occasion.

> Simultaneously, the floodlight system installed outside operated on the hall, church, parochial house and the factory building and the whole scene was one to be remembered. An illuminated address commemorating the event was presented to the Minister for Industry and Commerce through Mr Cosgrave… .
> The party then left for the newly opened Mineral Water Factory and after an inspection proceeded to the church for Benediction. Later, over one hundred guests sat down to a dinner in the Station Hotel where various speeches were made and toasts drunk to the bright future which would follow the coming of electricity to the village and its surrounding rural area.

Local merchants were among the first to appreciate and avail of the new power. In Tulsk, Co. Roscommon, one of the first of the newly developed fluorescent lights made its 'rural' bow in a small thatched public-house on the night of the switch-on in summer 1948. In September 1950 ice-cream was on sale in the village of Templetouhy, Co. Tipperary, one hour after the first consumer had been connected.

In the December 1951 issue of *REO News* there is a long poem of six stanzas entitled 'The Lighting of Scariff' by a member of the crew, Oscar Hannon, who sang it to the air of 'The Mountains of Mourne' at the switch-on celebrations. At least one stanza deserves quotation as illustrating the unsophisticated joy of such occasions:

> Oh Mary now Scariff's a wonderful sight
> With the people all getting the 'new fashioned light'
> After watching and waiting for many a year
> At least they've consented to send it out here.
> An badly 'twas needed as you must well know
> For in Scariff at night-time wherever you go
> There are dangerous angles all over the place
> Where many a craythur was stretched on his face.

Such 'poems' were not unusual. At the switch-on of Killanny, Co. Louth, a similar tribute to 'The Boys from the ESB' was rendered by a local schoolteacher, the last stanza of which went:

> Transformers they of rural life —
> A blessing on their head
> They give you heat, they give your power,
> They banish darkness dread
> They'll 'fluoresce' the old land
> From Finn to lovely Lee
> They know their job and do it well
> The boys from the ESB.

In some areas local pride was not content with a single ceremony. In the Ballinlough area on the Roscommon/Mayo border there were three separate switch-ons, the first was held in Ballinlough itself as befitted the importance of the principal centre. This was quickly followed by a second in Cloonfad and a third in Garranlahan, each occasion being marked by a supper and dance. Neither were food and drink stinted. In Drumlish, 'the local committee provided an excellent supper which included fresh salmon and champagne, followed by an excellent musical programme'. Champagne was also produced at the switch-on banquet in Woodford, Co. Galway, where afterwards two celebration dances were held simultaneously — one the rather sedate official function in the Parochial Hall, and the second a more boisterous affair organised by the crew members themselves in the commercial ballroom to which a half-barrel of Guinness had been rolled up the village street, the gift of a grateful publican.

In many northern areas where the proportion of Catholics and Protestants approached something like parity, the blessing and switching on of the supply were carried out jointly by the pastors of both denominations. In the predominantly Catholic south, however, the task was generally given to the parish priest, with the clergy of other denominations attending as guests of honour. In a midlands village, obviously not notable for its ecumenical leadership, the elderly parish priest was reputed to have settled an old score with the local C. of I. rector during the blessing ceremony. Having duly sprinkled the switch with holy water, he turned and also sprinkled the audience, with an extra special douche for the rector, who, denied a place on the platform by the parish priest, had been placed by the committee in a seat of honour in the middle of the front row.

The engineer in charge of the Clifden area recollected the switch-on ceremony in the local hall. In the centre of the stage was the ubiquitous large metal-clad switch which had given trojan service at many a switch-on and whose cover was 'liberally bespattered with rust spots from frequent former applications of Holy Water'. On the platform were both the parish priest and the C. of I. rector, but as representative of the majority faith the task of blessing and switching-on fell to the priest. All went well with the ceremony until the moment came to pull the switch. Amid a hush of expectancy, he operated the lever. Nothing happened. There was a shocked silence for some seconds and then a loud whisper came from the back of the hall: 'Let Rev Mr Fairbrother have a go!' As the tension broke and the audience exploded into laughter the reverend gentleman obliged by taking a bow. Immediately afterwards the hall blazed into light amid tremendous applause: a minor electrical fault had been rectified by an alert electrician backstage.

There were some very special switch-ons to mark special milestones. In February 1952 the fifth anniversary of the connection of the first rural consumer was celebrated when the 55,000th consumer, the parish hall in Kilsaran, Co. Louth, was switched on by the parish priest, Fr McEvoy, in the presence of a platform party which included Frank Aiken, then Minister for External Affairs, and the ESB Chairman, R. F. Browne. An exhibition of electrical equipment was mounted by REO and local traders. This included, for the first time, a television set as County Louth was on the fringe of the service area of the BBC's Northern Ireland transmitter. It is recorded in the area report, however, that 'unfortunately, or perhaps fortunately, it was a bad night for reception or he [the TV dealer] would have stolen the show'.

Even this event was eclipsed by the switching on of the 100,000th rural consumer, the Ballinamult Creamery of the Knockmeal Co-Operative Society in County Waterford, by the Minister for Industry and Commerce, Seán Lemass, on 1 March 1954. On this occasion, as well as a large contingent of ESB chiefs, the attendance included all the local TDs, senators, county councillors and other public representatives and – to quote *REO News* – 'all the residents within miles of Ballinamult'. Television continued to push itself to the fore as 'with the co-operation of Pye (Ireland) Ltd. the complete proceedings were televised with receivers placed to show the ceremony to overflow crowds both inside and outside the building'.

In May of the same year (1954) a very special ceremony took place in Rosmuc, Co. Galway, when the Taoiseach, Éamonn de Valera, attended the switch-on of the area which took place outside the former cottage of his old friend and comrade-in-arms, P. H. Pearse. It was an open-air occasion and a huge throng was present to watch the Taoiseach press the switch, which he hoped was symbolic of the better future for the country for which Pearse had dreamed, worked and died. It is notable that this was the only occasion on which de Valera as Taoiseach participated in such a ceremony.

Perhaps the most original approach to a switch-on was in Mountgordon in County Galway. At the request of Fr Hennelly, PP, a film of the event was made by the members of the camera club in Castlebar. Shots were also taken of the area crew at work during various stages of construction. The edited film was sent to America to play an important part in an appeal for funds for the building of a new church at Park.

Many hundreds of other switch-ons took place which did not make national headlines, but which were just as momentous for the areas concerned.

A continually recurring question, raised especially by members of the older generation, was whether the electricity now lighting up their locality really came from the Shannon Scheme. When assured that this was substantially so, although of course there were now other stations also contributing, their pride and joy were obvious. For them the Shannon Scheme had been one of the first great manifestations of an independent nation and on this night they were at last partaking of its fruits in their own community.

The construction crew also shared in the general euphoria of the occasion as an extract from an article in the final issue of *REO News* points out:

> Not all will perhaps admit it, but there were very few Rural Area Engineers who did not feel a glow of pride and satisfaction as they surveyed their first group of consumers after they had been switched on. This pride was also evident in the faces of the crew as they gathered, as was general in those early days, at the local church, hall or school and watched a local celebrity press a switch and listened to the gasps of delight as night was made into day; or as they strolled down a village street under the light of the new street lamps; or listened to the purr of a newly-installed electric motor carrying out its task of relieving some of the toil and drudgery of the country side.
>
> Then, the difficulties and squabbles receded into the background as the worthwhile nature of their work came to the fore. Theirs was the satisfaction of knowing that in their own way they were leaving this part of the world a better place than they had found it.

CHAPTER THIRTEEN

Threading the Way

THE FIRST POLE OF THE RURAL ELECTRIFICATION SCHEME was erected at Kilsallaghan, Co. Dublin, on 5 November 1946. It was a dull autumn day. There was very little ceremony. Present were W. F. Roe, P. J. Dowling and a small gathering of REO staff. It was to be a long journey to the erection of the millionth pole and the connection to the national electricity network of well over 400,000 rural households in 792 areas.

Government instructions were that one area in each county should be developed initially, so as to spread the benefits as widely as possible. This created some problems, as some counties were far better organised than others to initiate the preliminary canvassing necessary in order to have areas considered for selection.

The second area to be selected was Patrickswell, Co. Limerick, and work commenced in February 1947. This was followed in May by Inniscarra (Co. Cork), Tinryland (Co. Carlow), Multyfarnham (Co. Westmeath) and Easky (Co. Sligo). In July, Pollroan (Co. Kilkenny) was commenced, followed in September by Ballymacelligot (Co. Kerry), Ardrahan (Co. Galway) and Kilbride (Co. Wicklow).

In his monthly progress report for September 1947 the Engineer-in-Charge informed the Board that the first area – Kilsallaghan, Co. Dublin – was now completed and that construction was proceeding in nine other areas. In three further areas survey was well advanced and was commencing in seven others. Counties in which no area had yet been selected were Cavan, Clare, Donegal, Monaghan, Offaly, Roscommon and Longford. The year 1948 saw work starting in the remaining counties:

January: Kill (Co. Kildare), Murrisk (Co. Mayo)
February: Ballyduff (Co. Waterford), Bansha (Co. Tipperary), Carrigallen (Co. Leitrim), Clonaslee (Co. Laois)
March: Abbeyshrule (Co. Longford), Lusmagh (Co. Offaly)
April: Quin (Co. Clare), Termonfeckin (Co. Louth)
May: Mantua (Co. Roscommon)

June: Shelbourne (Co. Wexford)
August: Julianstown (Co. Meath), Rossnowlagh (Co. Donegal), Tydavnet (Co. Monaghan)
December: Gowna (Co. Cavan)

In most Districts, the first areas selected were used as training grounds for crews destined for work in later areas. For example, in Sligo District the structure of the crew that commenced work in Easky in May 1947 was developed in such a way that it could later split into two trained crews, one of which moved to Carrigallen in Leitrim and the other to Mantua in Co. Roscommon. As work proceeded in each area, the design and pegging of pole positions and other preliminary work went ahead in the area next on the list so that on changeover from a completed to a new area, the crew was able to swing into action with the minimum interruption in work flow.

A comparison between 1947/48 and 1948/49 shows the rapid acceleration in production achieved in the course of one year:

	1947/48	*1948/49*
Miles of line built in year	760	1,720
Number of poles erected	15,986	32,002
Number of consumers connected for year	2,227	9,353

By 31 March 1949 the ESB was able to report to the government that, despite the many difficulties in obtaining materials, the scheme was well under way. 49,000 poles had been erected and conductor strung on 2,490 miles of line. 11,500 rural households had been connected to the national electricity network in thirty-seven completed rural areas. Owing to escalation in material prices, the costs had been higher than originally anticipated and in the ESB Annual Report for the year 1948/49 the adequacy of the 50% capital subsidy was questioned: '... to allow the Scheme to proceed the Board obtained an undertaking that any deficiency during the first two years working would be covered by an additional subsidy from the Transition Development Fund. Negotiations were in progress to ascertain the additional sum required.' This emphasis on the inadequacy of the subsidy was repeated in the 1950 and 1951 Annual Reports. Nevertheless, there was no slackening of pace. Indeed, for the years ending 31 March 1951 to 1953, forty-nine areas were completed in each year. For those areas at the bottom of the queue, however, this was not fast enough. Pressure grew through their local representatives and in turn from the government to the ESB for an acceleration in the pace of working. In June 1953 it was decided to

increase the rate of development by 50%. Additional material was ordered, extra staff recruited and trained and the design and survey of new areas undertaken.

In its Annual Report for the year 1953/54, the Board was able to record that development had now reached the rate of seventy-five areas per year. It further indicated that a new target rate of one hundred areas a year had been set for the year 1955/56. Ninety-nine areas were in fact completed. It would be of interest to pause at this point, 31 March 1956, to review the progress in the first nine years. 499,000 poles had been erected, 43,000 kilometres of line strung and 163,000 consumers connected in 463 completed areas. Thus, after a difficult start, over one half of the proposed scheme had been completed. A further cause for satisfaction was that, after a slow start, the connection rate in developed areas now averaged 67% of all premises, which was almost up to the White Paper target of 69%. The capital subsidy had been withdrawn by the Interparty Government in October 1954, but this was not allowed to retard the pace of development.

Electric lighting as a substitute for the oil lamp was the first and most widespread use. REO staff vigorously advocated the installation of power sockets both in the house and in the out-offices to make it possible to reap the benefits of electrification in other ways. In the early years, however, this was a very slow process, especially in the case of the less affluent and less progressive. Major changes in life style could not be expected to take place overnight. In order to get some idea of progress in this field, REO conducted a survey during 1953 in a number of areas that had been connected for some time. Results were encouraging. As was to be expected, most progress had taken place in the home. Almost 80% had mains radio (the saving on battery-charging costs frequently went a long way to offsetting the fixed charge), 64% had electric irons, 44% had electric kettles and 25% had installed electric cookers. Washing machines and electric wash-boilers tied at 17%. Progress in installing piped water, one of the greatest benefits hoped for with the coming of electricity, had, however, been disappointing — only 11% had water pumps installed.

Out in the farmyard, development had been slow. While 25% had installed lights in the yard and in the out-offices to extend the winter working day, the number of power sockets installed and the out-dwellings was very small. About 9% of farmers were using electric motors for such chores as root-pulping and the same percentage had installed infra-red lamps for pig and chicken brooding. The electrification of milking was still in its infancy, with only 6% using electricity for this purpose.

In the case of the first seven areas, which had been connected in 1947 and which thus had had most time to adapt to the new resource, it was encouraging to note that the average annual electricity consumption per consumer, which in 1948/49 had been 509 units, had reached 1,145 by 1953/54. This average figure, however, concealed a wide variation between areas, ranging from 1,611 units per consumer for Patrickswell, Co. Limerick, to 625 for Ballymacelligot, Co. Kerry.

The peak of construction activity was reached in the years 1954/55 to 1956/57. For the year ended 31 March 1955, seventy-five rural areas were completed and 27,316 new consumers connected. In the following year, this performance was overshadowed when ninety-nine areas were completed and 34,257 new consumers added. In 1956/57 the number of areas recorded as completed was eighty, but the new consumers added reached the record total of 34,627. At this stage, almost 200,000 consumers out of the initial target of 280,000 had been connected at a total cost of £20,661,000. The rate of capital expenditure had also reached a high peak: for the two-year period of 1955/56 and 1956/57 it had exceeded £7.8m and was over 10% of total national capital expenditure. With the end of the initial phase now in sight, it was necessary to commence a controlled tapering down of activity to relieve the high demand on scarce national capital resources and to avoid a serious problem of staff redundancy, which a sudden stop would create. Thus, the number of construction crews, which had reached forty in the peak years, was gradually reduced and key personnel were absorbed into other activities. The number of new areas completed showed a corresponding phasing down, as seen in table 4.

TABLE 4

New Areas Completed and Consumers Connected

Year ending	*No. of new areas completed*	*No. to Date*	*No. of Consumers connected*	*No. to Date*
1958	60	603	17,725	215,243
1959	45	648	14,997	230,240
1960	40	688	13,458	243,698
1961	40	728	11,570	255,268
1962	26	754	9,515	264,783
1963	19	773	8,829[1]	273,612
1964	12	785	10,820[2]	284,432
1965	7	792	11,773[3]	296 ,205

[1] (including 3,800 'post development')
[2] (including 7,692 'post development')
[3] (including 10,776 'post development')

The 792 areas developed under the initial phase covered an area of 23,400 square miles or 86.5% of the country. If the areas of poor land were deducted (i.e. land with a Poor Law Valuation of 5 shillings an acre or less, into which the extension of the scheme was not contemplated in the White Paper) it could be claimed that 98% of the remainder had been covered at this stage.

It will be seen from the table, however, that a new phase had commenced in 1963 — the 'post development' phase. This involved the sending back of crews into already developed areas to connect householders who for various reasons had not accepted supply or who had been regarded as 'uneconomic' during the initial development of their areas, but who were now pressing for connection. The Electricity (Supply) Amendment Act 1962 enabled this to be done by permitting the ESB to extend the benefits of the scheme to areas hitherto classified as 'uneconomic' and on improved terms to unconnected houses in developed areas. Thus, connections under the original phase diminished and the number of 'post development' connections rose as the geographical spread of the scheme approached completion. By the end of the year 1964/65, at which point supply had been extended to all 792 areas, 296,000 rural consumers had been connected, of which 23,000 had been supplied under 'post development'.

While this signalled the completion of the scheme as envisaged in the White Paper (i.e. 280,000 consumers or 69% of the total dwellings), it was by no means the end of rural electrification. When the White Paper was published towards the end of the war years, the target figure of 280,000 or 69% of houses connected could be considered acceptable. By the 1960s, however, the 31% thereby excluded were in no way prepared to accept a future without electricity. After the stagnation of the fifties the economy was beginning to pick up and the benefits of electricity were now becoming manifest even to the most conservative. Intense pressure grew for a fresh look at the conditions for subsidised connection of electricity supply.

A 'planned post-development' (PPD) scheme was agreed between the ESB and the government whereby every householder in the 792 areas who was still without electricity would be given a further opportunity to obtain supply on subsidised terms. Between 1965 and 1971, 50,000 extra consumers were connected under this scheme, so that by 31 March 1971 the number of 'rural' consumers stood at 346,000.

Even at this stage, however, when close on 90% of all rural homes had now been supplied under the scheme, there was no respite for the ESB. The clamour from the remaining householders, mostly located in isolated pockets, grew in intensity and the pressure by their local representatives on the government and by the government on the ESB was irresistable. A further post-development scheme was agreed under which the Board undertook to offer every unconnected householder in the 792 developed areas a last chance of obtaining a subsidised connection. Any new connections after this would be on a strictly commercial basis, without any help from government subsidy (See chapter 20 for a more detailed account of these schemes.) About 60,000 consumers were connected under this PPD scheme. The Board in its 1975/76 Annual Report recorded that by 31 March 1976 the total number of consumers connected under the Rural Electrification Scheme was 405,890, or around 98%/99% of all rural dwellings. Only a few thousand houses situated in the most isolated regions and, consequently, most expensive to service, now remained without supply. The initial concept of a 69% connection as being all the economy could bear had, with the introduction of the various post-development schemes, long been thrown out the window. Now electricity supply, even to the most isolated dwelling, began to be portrayed as the democratic right of every citizen. While most of the still unconnected householders were scattered in ones or twos in remote situations all over the country, in some locations they were sufficiently grouped to form strong pressure groups. Two examples were the Black Valley, a very isolated and scattered community of about forty houses south of the Gap of Dunloe in the Macgillycuddy Reeks area of County Kerry, and Ballycroy, an almost equally isolated group in western Mayo. Strong and articulate representations from such groups persuaded the government to introduce special legislation at the end of 1976 to enable supply to be extended on a subsidised basis to about 1,600 remote dwellings in similar situations. This legislation effectively brought to an end the long series of extensions to the original Electricity (Supply) (Amendment) Act of 1944 which had had the effect of transforming the limited scheme originally envisaged into effectively the complete electrification of the country.

CHAPTER FOURTEEN

Spreading the Message

IN ORDER THAT THE SCHEME should be a success, both economically and socially, it was essential that the new consumers should not only be shown how to make the best use of the new power but should be actively encouraged to do so. This involved REO in a widespread programme of education, demonstration and promotion of the many ways in which electricity could help in the home and on the farm. To the majority, electrification meant the coming of the 'light' and this was of course the first and prime application in all households; but it had to be demonstrated that this was not all, and that there were many other ways in which electricity could be used to advantage.

Tremendous benefits would be reaped at low cost by the provision of an outside light to enable yard work to be carried out during the hours of darkness. This could be extended comparatively cheaply and with great effectiveness to the lighting of the farm sheds and byres, thereby lengthening the farmer's working day. Even this took a long time to be widely adopted.

In the house, the transition from soft but poor illumination by the traditional oil lamp to the brighter but harder electric light was frequently difficult to accept. All too often a low-wattage bulb was installed in the kitchen – the family living-room – both to minimise the contrast with the old oil lamp and to economise in the use of electricity which still had the image of an expensive alternative. In an effort to combat this reluctance, REO gave a free 100 watt bulb to every newly connected consumer for installation in the kitchen. It was hoped that after a short time the family would become used to a good level of illumination and would be reluctant to settle for less. This was effective only to some degree. In many cases the transition was too traumatic and lower-wattage bulbs, usually 40 or 60 watt, were quickly substituted. Even these were much brighter than the oil lamp. As late as 1958 in the Mount Delvin Area in County Galway the number of householders who substituted the smaller bulb was so large – about 60% of all – as to prompt the area staff to carry out a mini-survey as to the reasons. They were told that the 100 watt bulb was — too hard on the eyes, made the people feel sick or that the bright light tended to put out the fire. This reflected the widely held belief that any bright light, especially the sun, if allowed to shine on the traditional turf fire would tend to extinguish it. This was of course because it masked the glow otherwise seen in the comparative gloom of the hearthside. Another but seldom admitted cause of the dissatisfaction with the high-wattage bulb was that it was merciless in revealing dirt, dust, cobwebs, defects in the furniture, all crying out for higher standards of housekeeping.

REO News of March 1959 quotes the Rural Area Engineer in Carrigatogher Area in County Tipperary, who found that if he waited for a few weeks after connection before giving out the free 100 watt bulb, very few went back to a lower wattage. By then they had got used to the 60 watt level of lighting and so the jump to 100 watt was not so great.

The promotion of high-wattage bulbs for lighting was of course only scratching the surface of the development problem. Because of the cost involved, householders tended to confine the internal wiring of their premises to the minimum and the provision of adequate sockets was a very low priority.

In the early years the average cost of installation by rural wiring contractors was from £1 to £2 per point, but even this was considered beyond their means by many. In the Ballycastle area in north Mayo, the parish priest of Lackan, Fr Murphy, organised a bulk-buying co-operative which provided the wiring materials for about twelve shillings per point, the actual installation being a do-it-yourself job. Indeed much of the wiring in the more remote areas was on this basis or else was carried out by the local handyman. The circuits wired, however, were of the most basic type and did not require any great knowledge of electricity. Provided standard wiring materials were used it was difficult to make a mistake. Even so, requests for aid arrived regularly and a letter from a County Sligo consumer quoted in *REO News* (September 1957) ran: 'Dear Sir, Please send me some booklets and leaflets on how to wire houses and farmyards for the electric current, also send me some leaflets on how to treat a person for electric shock, write soon, yours truly. . .'. In the February 1954 issue came a story – surely apocryphal – from the Kingdom of Kerry where a 'contractor' extending the existing wiring in a rural dwelling was not quite sure if the earth connection was connected correctly and for safety tested it on the cat – 'the unfortunate animal entered the feline Valhalla by the shortest route'. The circuit fuse was then removed and, to make doubly sure, a cork was screwed into the fuse base pending the quick arrival of an ESB electrician.

REO promoted the use of electricity on two main fronts – domestic and farmyard. Much effort went into persuading householders to fit an adequate number of socket outlets. It was a difficult task and even in 1964, surveys showed that the average number of outlets at two and a half per premises. By 1974, however, following a decade of relative affluence, the average had risen to five and a half.

In a typical rural area the promotion of the use of electricity usually consisted of a series of stages as outlined below.

Preliminary Contact □ During the official canvass the Area Organiser who called at each house was able, with the help of suitable leaflets, to outline many of the benefits electricity could bring.

Direct Approach □ When the area was selected for development, the Area Organiser again moved in, visiting every house to check up on potential consumers. He raised the question of adequate wiring and went on to discuss suitable appliances. This was the first discussion at which it was clear that supply could be given and that the householder would take it.

Display □ Where possible, the Area Organiser secured a small showroom or shop window incorporated in the vicinity of the area office, where prospective consumers could view the appliances. Sometimes, as in Blackwater, Co. Wexford, the opportunity of a vacant shop was eagerly grasped for this purpose.

Good example □ The Area Engineer tried to have his 'switch-on' and to get consumers connected as early as possible in the construction period and the newly-delivered appliances immediately put to work. Those yet to be connected, or who had not yet decided to purchase appliances, were thus provided with practical examples of their benefits.

Demonstration □ At a convenient time during the construction period, local halls were hired for a demonstration week. This was a joint effort between the Area Organiser and a demonstrator from headquarters, who came with a large van fitted out with a comparative range of appliances and demonstration equipment. These vans were constructed so that in addition to carrying equipment for indoor demonstrations, they could also act as open-air demonstration theatres at country shows or other such functions. During the demonstration week the exhibition

Right: The direct approach. *Below*: Spreading the message.

was open during the afternoon, while at night a comprehensive programme, including films of domestic and farmyard electrical appliances, was staged, usually to a packed house. This programme also included practical electric cooking by demonstrators from the local District headquarters, the cooked dishes being shared out among the audience. At other times during the week, special visits were made by the Area Organiser and the visiting demonstrator to individual consumers – blacksmiths, small industries, etc. — who had special requirements outside the normal experience of the Area Organiser.

To combat its high-cost image the actual amount of electricity used in the various applications was measured by a meter and the cost computed. Having explained the system of charging for the various blocks of units consumed (in 1951, for example, the first eighty units per bimensal period cost 2.8*d.* per unit; the next 280, 1.3*d.* per unit and any units above 360 per period cost 1.05*d.* per unit), the demonstrator then indicated the amount of work a unit could do. It would light the 100 watt bulb for ten hours, do the ironing for a week, wash the clothes for a month (people were invited to bring bundles of soiled clothes), cook all meals for one person for one day and boil 18 pints of water. A member of the audience was invited to check the meter before and after to verify the claimed figures. In Kilvine near Claremorris in County Mayo a schoolboy checked that boiling a full six-pint kettle used 0.3 units. When then asked by the demonstrator how many kettlefuls a unit could boil, he reflected a moment and in a loud proud voice proclaimed 'three and a sup'. He then went on to calculate that this would provide eighty cups of tea.

These practical demonstrations did much to break down prejudices regarding the cost of electricity. For farmyard applications the demonstrators stressed the almost unbelievably low cost of electric motive power. The ¼hp motor used only one unit for four hours working, in which time it could perform better than the strongest worker. Using electric motors one unit would grind one hundredweight of meal, pump 1,000 gallons of water from the average well and milk one cow twice a day for ten days. At one venue the operation of a root pulper by the tiny ¼hp motor was being demonstrated to a cautious farmer (two tons pulped for one unit), when the farmer's young son was observed trying to persuade his father to invest and simultaneously urging the REO man to intensify his efforts when Dad appeared to hesitate. Evidently his interest sprang from the fact that he was responsible for pulping on the farm.

Demonstrations such as these were useful but the real task was to get consumers actually to purchase and install applicances. The success of a demonstration was measured by the sales which resulted.

The mobile showrooms.

Post-Development □ In many areas there were active electrical retailers – very often the same men who had wired the houses – who canvassed for business with the full co-operation of the REO staff. The first appliance to be installed was generally a mains radio, as the saving on the cost of battery-charging often went a long way to offsetting the fixed charge. The ESB did not enter into this field, as it considered that it was best left to the local dealer who was in a position to offer the specialist service required. It did, however, continue to remain active in the sale of domestic and farmyard appliances, as it was felt that the private sector was not equipped to provide the drive or the geographical spread necessary to ensure sufficiently rapid development. A fleet of well-equipped sales-vans was established, each van fitted out as a miniature display. The salesmen were usually drawn from the ranks of Area Organisers who had shown a flair for this work. An important appliance in their stock was the electric water pump and they were trained to answer most of the queries regarding the provision of water on tap. They also had a stock of domestic equipment which they could demonstrate in the actual farmhouse kitchen and leave for a few days trial if requested. The same applied to the farmyard appliances – electric motors, grain grinders, oat rollers etc. As the milking machine market was served by specialist firms, REO did not sell these machines, but availed of every suitable opportunity to demonstrate the merits of machine milking, particularly at agricultural shows throughout the country.

The numbers of the different appliances sold by the vans in the six-month period April-September 1953 and 1954 indicate how the market took off. The order of preference of appliances is shown in brackets.

	1953		*1954*	
Irons	160	(2)	479	(1)
Kettles	134	(4)	474	(2)
Cookers	211	(1)	419	(3)
Washing machines	111	(5)	164	(5)
Wash boilers	67	(6)	86	(6)
Fires	51	(7)	75	(7)
Vacuum cleaners	28	(9)	44	(11)
Refrigerators	13	(12)	26	(12)
Pumps	156	(3)	330	(4)
Grain grinders	20	(11)	51	(9)
Motors	35	(8)	61	(8)
Welders	25	(10)	50	(10)

The mobile salesmen worked closely with the local ESB office and showrooms, generally using it as a base for restocking and frequently for completing sales initiated in the farmyard or farmhouse. The local staff co-operated both in the actual selling and in the after-sales service. On the 23 July 1960 a wedding anniversary present of a washing machine, complete with appropriate greeting card and tied up in pink ribbon, was delivered punctually by the Kells staff to a very delighted woman out the country. It was from her husband then working in Nicaragua, who had conducted the whole operation by correspondence. A farm family near Kenmare, Co. Kerry, purchased an electric cooker which they asked to have installed in time for a 'Station' being held in the house a few days later[1]. This was arranged by the Area Electrician with two days to spare. The family was so pleased that he was invited to the function. To continue the story in his own words, 'the dear old lady had forgotten everything I had told her about the controls due to the excitement. She came to me in a state of desperation — would I ever show her again. Too late, I considered, she was beyond understanding anything I said by this time so I donned an apron, secured pots, pans, eggs, bread for toasting, etc. and I cooked the breakfast for twenty people, including two priests, clerks, farmers with big appetites and even for the people of the house (it was a Friday morning — hence the eggs).'

In the dairying areas the creamery was a great meeting place. Each morning the carts with their churns lined up and gossip was exchanged while the milk was taken in. This was duly noted by the REO staff and many early morning creamery demonstrations were held in the open air with gratifying sales results. A typical example recorded was at Upperchurch, Co. Tipperary, where from 7am to 9am on a sunny summer morning in 1957 a comprehensive demonstration of equipment was given, including electric cooking. A breakfast of sausages and tea was served by the AO. In Lislynn area on the Cavan/Meath border what was described as an 'Intimate open-air demonstration' was held for a small group of women — who were invited to bring along their washing! While the washing was being done, the merits of water on tap, electric water-heating and of course washing machines were discussed.

On occasions, an opportunity for extra publicity for the rural scheme was perceived, as in the case of An Rás Tailteann, the big national multi-stage cycle race organised in 1961 by the National Cycling Association of Ireland. One of the large rural demonstration vans, decorated with a suitable slogan, was employed as a mobile laundry for the competitors' clothes, using domestic electric boilers, washing machines and tumbler dryers. The thousand-mile race was run in eight stages from Friday to Friday, commencing and finishing in Dublin. The overnight stops were at Navan, Castlebar, Tuam, Castleisland, Killarney, Clonakilty and Wexford. The ESB van accompanied the competitors over the whole route and at each of the stops it was set up, opened out and up to five hundred items of soiled clothing – team jerseys, singlets, shorts, socks, caps etc. – were washed and dried in full view of the large crowds and were ready to hand back to the competitors next morning. The slogan on the van was 'Electricity Leads in the Race for Cleanliness'.

REO produced an extensive range of literature in the form of leaflets and booklets. This covered every phase of development, Rates of Charge, Farmyard Wiring, What a Unit Can Do, How Units Can Help, Domestic and Farm Appliances, Kitchen Planning, Water Pumping, Water Heating, Electricity for Poultry, The Electric Grain Dryer, Animal Food Cooking, Soil Warming. As a back-up to all these efforts, a continuous series of display advertisements was inserted in all fifty-two local papers, in the farming weeklies, in six rural and about thirty other weekly or monthly magazines. While these mentioned domestic equipment, they mostly laid stress on farming applications (Fig. 1), frequently quoting actual farmers' experiences. Domestic applications were in the main covered by weekly advertisements in the four daily and four evening papers. These were aimed at urban as well as rural domestic consumers, as by the end of the fifties the difference between the two in the use of electricity for domestic purposes was beginning to disappear.

Indeed, the arrival of the sixties marked a major break-through for the promotional efforts of the REO staff. On the broad economic front the stagnation of the forties and fifties was yielding to a more dynamic phase. In rural homes the REO promotional drive was now being carried out in the context of an improving economic situation and a growing receptivity to innovation. This was reflected in a growing increase in the use of electricity and a greater readiness to invest in electrical appliances, particularly in the domestic area.

The gap between urban and rural living standards began to close rapidly. More new houses were built and old ones modernised. The traditional dark rural kitchen became transformed, and the drudgery of rural housekeeping was relieved by the installation of hot and cold running water and labour-saving appliances. Table 5 traces the growth of appliance ownership among rural electricity consumers. By 1958, after a decade of rural electrification during which the national economy remained in low gear, the only domestic appliances that had been widely accepted were the low-cost iron and electric kettle. By 1968, however, electric cooking had moved in, as had the refrigerator, which made it possible to store perishable food and reduce the frequency of journeys to the local shopping town. Television, one of the greatest contributors

An early-morning creamery demonstration — and a breakfast of sausages and tea served by the AO.

to relieving the isolation of the countryside, had been eagerly adopted and after a further decade would be in almost every house. The high level of ownership of washing machines (59% by 1979) indicates higher living standards with less toil. In the early seventies home freezing cabinets appeared, reaching 18% ownership by the end of the decade. They permitted a higher quality and greater variety of diet. Their most popular use was in the long-term storage of home-killed meat.

TABLE 5

Domestic Applicance Ownership among Rural Electricity Consumers
(Percentage ownership rounded off to nearest 1%)

	Percentage of Consumers					
Appliance	*1958*	*1964*	*1966*	*1968*	*1973*	*1979*
Main Cooking						
Electric	2	13	15	17	15	25
Bottled gas	4	14	18	24	33	34
Solid fuel	84	70	65	57	48	35
Oil	10	3	2	2	4	6
Electric water heater	4	5	8	8	8	24
Electric blanket	1	4	7	12	20	35
Foodmixer	—	2	5	NA	18	38
Hairdryer	1	—	7	NA	17	48
Electric iron	68	79	82	85	86	87
Electric kettle	39	47	50	53	55	69
Refrigerator	4	8	14	19	44	76
Freezer	—	—	—	—	4	18
Television	—	23	43	55	NA	84
Toaster	3	4	5	NA	11	25
Vacuum cleaner	9	10	12	16	21	45
Washing machine	11	19	26	30	41	59

Source: ESB Time — Series Surveys, May 1981
NA: Not available

It is of interest to examine the growth of electricity consumption over the same period. Coincident with the increase in electrical appliance ownership was an increase in unit (kWh) consumption. Table 6 shows the distribution of this consumption. In 1964, 67% of rural consumers used less than 800 units for the year (21% used less than 200 units), while in 1979, 68% used over 800 units (27% used over 4,000 units). These figures are, however, national averages. As with many other consumption patterns in rural Ireland, the gap between east and west shows up in further analysis. A study by Michael A. Poole of Queen's University Belfast for the year 1965/66[2] shows a definite falling off in electricity expenditure per rural household as one moves from the south-east towards the north-west, the average expenditure east of a line from Dundalk to Tralee reaching double that of the area west of this. Another interesting indication from Poole's study was that variation in electricity expenditure was more closely correlated with local bog availability (perhaps a source of alternative fuel) than with income.

TABLE 6

Annual Electricity Consumption Per Household Among Rural Electricity Consumers

(Rounded off to nearest 1%)

Electricity consumption	*1964*	*1966*	*1968*	*1973*	*1979*
	Percentage of consumers				
Up to 200 Units (kWh)	21	17	14	11	8
200 – 400 kWh	23	17	14	10	4
400 – 800 kWh	23	22	22	20	8
800 – 1200 kWh	9	11	12	15	11
1200 – 2000 kWh	10	12	14	15	18
2000 – 4000 kWh	9	12	15	15	23
Over 4000 Units	5	8	10	12	27

Source: ESB Time-Series Surveys, May 1981

CHAPTER FIFTEEN

Exhibitions and Competitions

THE RDS SHOWS

IN 1946, the year the Rural Electrification Scheme was launched, the Royal Dublin Society had been 215 years in existence. It had been founded in 1731 for improving 'Husbandry, Manufactures and other Useful Arts and Sciences'. In its later years much of its good work was done through its Shows. The first livestock show and exhibition of agricultural implements by the Society was held in April 1831 in the yard adjoining Leinster House and the first Spring Show was held in Ballsbridge in April 1881. From then on it was held almost every year, interrupted only by the two World Wars and the War of Independence, and it grew steadily in status and appeal, especially to the farming community.

The Spring Show was first attenuated and finally suspended completely during the 1939–45 war and was resumed in 1946. Attendances boomed as farmers and their families, isolated and limited in mobility during the war, were now anxious to catch up with the latest developments in agriculture and agricultural equipment.

The value of the Show as a vehicle for communicating with the rural community was quickly recognised by the Rural Electrification Office and it approached the RDS for suitable facilities to mount an exhibit. The RDS in its turn, recognising the potential contribution of electrification to the improvement of Irish farming and farming life, enthusiastically welcomed REO and co-operated whole-heartedly in providing suitable exhibition space.

From 1947, when the first Rural Electrification stand took its bow, up to the time of writing, the Rural Electrification Exhibit has been a regular feature of the RDS Spring Show. For a few years, exhibits were also mounted at the Horse Show held in August. The Spring Show attendance, however, was far more representative of the wider rural community, and it was to this Show that most attention and resources were dedicated.

In the late forties, when the pay-back from farming was low and the value of electricity and electrical aids to the majority of rural dwellers had yet to be proven, the emphasis was on the introduction of the benefits of electricity on the farm in a simple and inexpensive way to unsophisticated and inexperienced audiences. Year by year, as markets improved and as farming became more specialised and market-orientated, the REO exhibits changed accordingly. The electrical appliances and equipment shown on the stands were more and more designed to meet specific farming problems and were subjected to critical appraisal from increasingly knowledgeable farmers and farm wives. The changing exhibits reflected in a striking way the social, economic and technological revolution taking place in rural Ireland.

The aim of the early exhibits was to demonstrate the many ways in which electricity could be of value, not alone in the home but out in the farmyard. Great emphasis was put on the contribution which could be made by a small and inexpensive electric motor in relieving the farmer and the farm housekeeper of many of the time-consuming chores which beset the ordinary rural homestead. The low cost of its use was dramatised in many ways to an audience pre-conditioned to believe that electricity was an expensive commodity to be used with great circumspection. In particular the ¼hp single phase electric motor costing about £10 was highlighted as being capable of doing a host of jobs around the farmyard such as root-pulping, sawing firewood, operating a grindstone or bench drill, water-pumping or, in the farm kitchen or dairy, churning or separating, all at a cost of one farthing's worth of electricity per hour, twenty-four hours a day if need be.

In order to convince the sceptics recourse was had to many dramatic demonstrations of this tiny machine. The most successful, in the 1950 Show, was the coupling of a small generator to a converted bicycle so that a volunteer from the audience could compete against the ¼hp motor over two minutes by pedalling as furiously as possible for that period. The pedaller's efforts were recorded on a dial in the form of 'watt minutes'. At the end of the period most volunteers were exhausted and chagrined to see the little motor calmly exceeding their best efforts and prepared to go on doing so indefinitely. One competitor of herculean physique, by exerting a tremendous effort, to the cheers and encouragement of the large audience, actually succeeded in pushing the pointer up to slightly better than the motor before almost collapsing from exhaustion. He was unique, however, and his nearest rivals were left way behind. None of those who had a go, nor indeed their audience, needed further convincing of the capabilities of the small motor.

From the first exhibit onwards, the value of running water on the farm and in the home was stressed as was the comparative ease and low cost with which this could be done. The farmers' penchant for doing as much as possible with their own hands was recognised, and in the 1949 Show three systems for obtaining running water at the kitchen sink were shown, all of which were capable of installation by the average handyman. The cheapest and most basic was a small centrifugal pump operated by the ubiquitous $\frac{1}{4}$hp motor, discharging through a $\frac{1}{2}''$ pipe into the kitchen sink. Not even a tap was needed, as a cord switch over the sink started and stopped the motor. A more ambitious installation, which could provide the foundation for a hot water system, was a self-contained centrifugal pump feeding a roof tank and controlled by a float switch. Finally there was the 35-gallon capacity pressure-storage pump unit which required no roof tank. This was shown feeding a complete house system including an 11-gallon water-heater and milk-cooler.

As time went on the water systems shown on the stand got more elaborate and more specialised. Emphasis was put on the self-contained pressure storage system which provided excellent pressure (20 to 40 lbs per square inch was the most common range), and which could be installed with the minimum of structural work. To feed the pressure-storage system a range of pumps was shown to meet the various well conditions and water requirements. For shallow wells, the simple centrifugal pump or the piston pump, usually belt-driven by a $\frac{1}{4}$hp motor; for deeper wells, the centrifugal jet pump; and for the deepest, the piston operated deep well pump and, later, the submersible pump, capable of insertion down a narrow bore hole to an almost unlimited depth.

Various other aspects of the provision of running water were highlighted. The grants available from the Departments of Agriculture and Local Government got full publicity. In order to achieve impact, competitions were held, as in the case of the giant tap delivering around 4,000 gallons per hour, where competitors were invited to estimate the amount delivered over the Show period. 'Gimmicks' were employed to excite interest — the Magic Tap from which water poured continuously without any apparent connection to supply, and the Mystic Kettle, apparently suspended from a rope without any electrical connection but which boiled continuously.

Group water schemes were also given publicity, the most elaborate exhibit being the huge model of the Kilcornan Group Scheme shown at the 1962 Show.

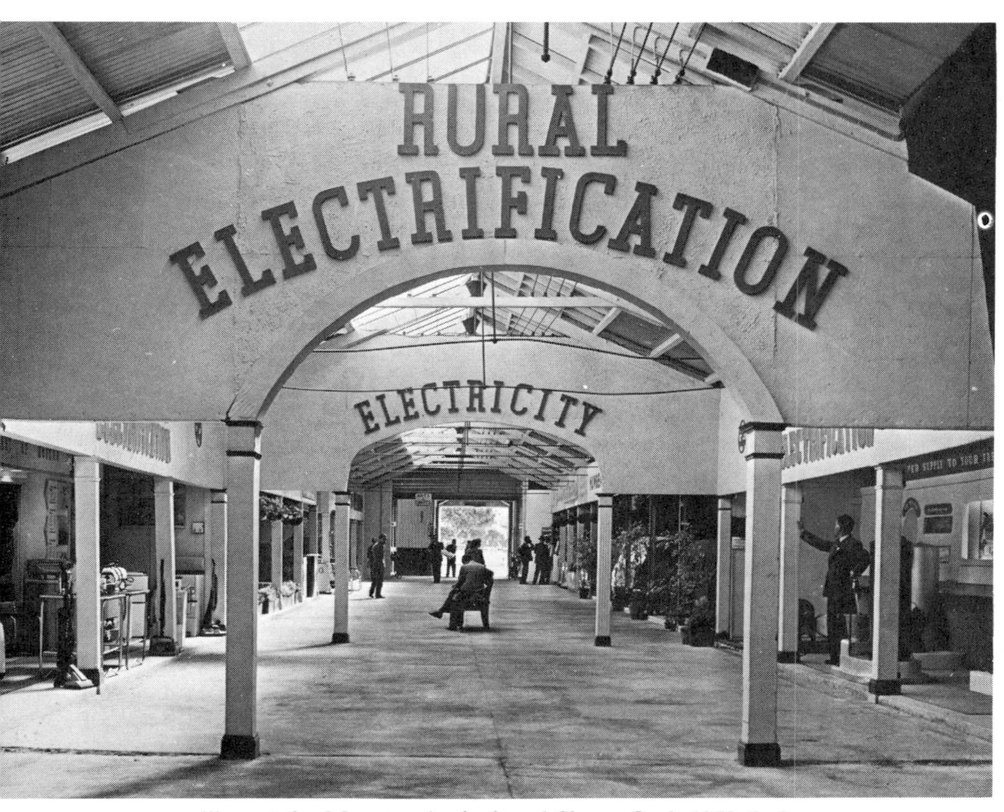

ESB pavilion at the Munster Agricultural Show, Cork 1952, before the onrush of the crowds.

Water-heating also got its share of promotion as various appliances for heating water in the home and in the dairy were shown. Year after year the message was repeated: 'Water in the home for cleanliness and comfort; water in the farmyard and on the farm for clean conditions, less drudgery, more profit; water, hot and cold, in the milking parlour and the dairy for hygiene and quality milk'.

In the early post-war years milking machines were not yet universal in the dairying counties, and one of the greatest sources of drudgery was the twice-daily seven-days-a-week milking chore. Most farmers with large dairy herds looked to rural electrification to provide a solution to the problem, and electric milking was featured in increasingly sophisticated forms in successive rural electrification exhibits. Owners of smaller herds tended to be more cautious. In 1948 a Canadian portable milking machine for the small herd owner was shown costing £54.6*s*.4*d*. for the single-bucket and £80 for the two-bucket model. Following assertions that 'cows would kick at the machine', REO's Publicity Officer, Paddy Ennis, offered to milk two cows which had never been machine-milked before, simultaneously, with the two-bucket version, in any location where electricity was available. The challenge was taken up by a North Cork farmer and down on the farm, before a large audience, the successful milking of the two cows was achieved in four minutes. The cows were completely unruffled but the operator was under some tension. It was only when the applause had died down that he admitted that he had never milked a cow in his life before, either by hand or by machine. As it turned out this was the most telling point of the whole demonstration: neither the cows nor the operator required previous experience!

A visitor from India to the stand had never seen a milking machine. He tested the vacuum pull on the cups and asked if it could be made stronger. He was shown how this could be arranged and departed well satisfied remarking that the original fifteen inches of vacuum would not be a strong enough pull for use on a buffalo!

The opening of the RDS's new Simmonscourt extension to Spring Show exhibitors in 1957 provided REO with an opportunity to stage a more elaborate exhibit. The RDS offered a spacious site in a choice location. Full advantage was taken and a full size model farmyard was built which was to become known generally as the 'ESB Simmonscourt Farm'. On the 'farm' there was a full display – usually of working examples, which the farmer could try out to his own satisfaction – of the various electrical aids to farming: grain handling, featuring augers and elevators; grain drying, showing at various times different types of dryers with suggested loading and unloading arrangements; provender milling with working examples of hammer mills, plate mills, stone grinders, oat rollers and food-mixers with practical examples of mounting and feeding arrangements; milking machines from the portable machine to in-stall milking and the milking parlour with full-scale layout. On some occasions a herd of pedigree milch cows was exhibited which were milked twice daily. Manufacturers of farm buildings and suppliers of equipment co-operated gladly in providing the ever-changing framework for the exhibit and reaped the benefits of the resulting publicity. The layout and equipment were changed regularly to keep pace with developing technology. Close co-operation was maintained with the Department of Agriculture, An Foras Talúntais, Macra na Feirme and other authorities to ensure authenticity. On at least one occasion the layout was designed by members of Macra na Feirme, whose members staffed the stand, to answer the multitude of queries on the farming practices illustrated. In 1961 there was a pig-rearing display which was a replica of a carefully costed project, carried out in Coláiste Charman, Gorey.

On the occasion of the 1969 Show an elaborate 'living exhibit' was staged which attracted great attention. To quote the *Irish Farmers' Journal:*

> Among the talking points of the Spring Show, the dairy layout in the Simmonscourt Extension is surely a front runner. Two aspects of it should be underlined. It is on the ESB stand, it is designed and staffed by the Department of Agriculture and it was built by the Irish Sugar Co. In this way it is an indication of what can be done when State and Semi-State resources are rationally pooled. It is also a living exhibit in a show that has too many dead ones. A herd of Friesian cows from Warrenstown (more co-operation) can be seen doing what cows do in kennel conditions, and there is actual milking carried out to show how quality bonuses may be earned. In short there is a good index on offer and farmers are given a fair chance to assess it. They are not left with a taste of it and told to use their imagination for the rest.

This particular exhibit was filmed by RTE and featured in their 'On the Land' television programme.

Advantage was also taken of the extra space in the Simmoncourt extension to show, at the 1957 Show, the first version of the ICA/ESB Farm Kitchen, designed by Eleanor Butler and staffed by members of the ICA as well as the ESB. The application of electricity to horticulture also made its bow. Electrical heating, soil warming, soil sterilisation, plant irradiation, bench warming, micro climate and mist propagation were demonstrated under actual working conditions. As the technology of the horticultural industry developed this was reflected in the horticultural exhibit, especially in the environmental control area with plant irradiation and 'growing rooms' in which the total environment of plants was precisely regulated.

In 1960 came a radical development. The Farm Kitchen had aroused great interest among rural housewives at previous shows, but of necessity it had initially been housed in an unprepossessing prefabricated building. Now it was decided to build the Farmhouse of the Sixties. As the economy began to take up after the stagnation of the fifties, new houses were being built in ever-increasing numbers in the rural areas. REO in conjunction with the Agricultural Institute and its architect, Paddy Tuite, decided to exhibit a carefully designed prototype farmhouse which would break completely from tradition and incorporate the best aspects of design, not alone in the kitchen but throughout the house as a whole. It would be specially designed to cope with farm life. It recognised that the normal method of access was from the back, directly from the farmyard, and that the front door was little used except for special visitors. Great attention was therefore paid to the back entrance, directly off which was a parking place for muddy boots and a shower room, wash basin and toilet. The adjacent kitchen was large enough to serve as the normal living area for the farmer and the family, and was of course equipped with a wide range of electrical aids. Adjoining was a utility room where chores such as laundry and ironing could be carried out without interfering with the activities in the kitchen. Here also were drying facilities for the wet outdoor clothes which are an occupational hazard of farming. The Farmhouse was an instant success. The *Irish Farmers' Journal* had this to say:

> The ESB stand in the Simmonscourt Extension has, for the past few years, been one of the most ambitious at the Spring Show. This year it is more ambitious than ever. The farmhouse, so long a focus of interest for rural housewives, has been completely re-designed with the co-operation of the Agricultural Institute and is now a fully furnished and equipped three-bedroomed house. Anyone thinking of building a new house ought to see it.

A small indication of rising standards was the installation and complete acceptance in 1961 of a wash-hand basin in the principal bedroom in addition to the normal fittings in the bathroom. Some years earlier this would have been derided as a pretentious luxury. In this year also a TV set made its appearance in the living room.

The deep-freeze cabinet first appeared on an exhibit in 1960 when pupils of St Martha's College, Navan, were invited to give a display of

Right: The ESB/ICA farm kitchen at the 1957 Spring Show. *Below*: Interested onlookers at grooming demonstration.

cheese making and processing of broiler chickens for the table. The processed birds were then deep frozen for market. The following year the girls were again invited back to the stand to demonstrate the processing and storage of broiler turkeys. Later in the decade the deep-freeze appeared as part of the kitchen equipment. The economic and dietary advantages of being able to store a whole animal carcass in prime condition had begun to appeal to more and more farm families. Thus in 1972 at the ESB stand we find Mr Schwer of the Meat Research Section of An Foras Talúntais demonstrating to rapt audiences how to reduce quarters of beef to cuts suitable for the freezer.

In 1973 Ireland entered the EEC and the markets of Europe were open to the Irish farmer. The stagnation and lethargy of the fifties were being left behind as the industry took off into a new prosperity. This was in turn reflected in the exhibits at the ESB stand. Specific requirements for achieving quality and quantity production were identified. Instead of simply promoting hot water in the dairy, the stand now exhibited 'A dual-purpose water-heater manufactured by an Irish firm to provide water for udder-washing and the once a month (170°F) hot wash recommended by An Foras Talúntais for dairy farmers using chemical cleaning'. 'Wet mix' and 'dry' feed systems for large pig enterprises, slurry disposal installations and water treatment for the removal of iron and excess minerals were now among the items listed on the stand.

An awareness of the necessity for environmental protection was indicated by an exhibit dealing with research into more effective effluent control. Electricity's contribution to land reclamation was shown by exhibits dealing with pumped drainage. (Two million acres of swamp land could benefit thus.) The range of electrical farming equipment requiring three-phase supply was reflected in the showing of single-phase to three-phase converters. The build-up of native industry was illustrated by the fact that these converters and indeed much of the equipment on show were now of Irish manufacture.

In addition to being prominently featured at the RDS, rural electrification was also promoted vigorously at all the main agricultural shows throughout the country. After the Spring Show the most elaborate exhibit was at the Munster Agricultural Show held in Cork in early summer, but other important venues such as Limerick, Tralee and Athy were also given special attention. Very frequently modifications of Spring Show exhibits were staged, in many cases following up a theme such as water supply, grain handling, processing of feed etc. At the smaller shows the mobile demonstration vehicles were the backbone of the REO exhibit, sometimes backed up with additional features. Many of the patrons of the country shows would have already been to the Spring

Show and thus it was quite common that a discussion on some application of electricity to farming initiated at Ballsbridge might be continued with REO staff on 'home ground'.

THE NATIONAL WHOLEMEAL BREADMAKING COMPETITIONS

The concept of national competitions for the baking of traditional Irish home-made wholemeal bread arose from an exercise in co-operation between the National Ploughing Association and REO on the occasion of the World Ploughing Championships held in Killarney in 1952. On this occasion the NPA requested assistance from REO in providing electricity to power various trade exhibits on a site remote from any mains electricity network. REO responded by lending and operating a number of its mobile generators. Apart from helping out the NPA, it saw in the huge and mainly rural attendance an excellent opportunity to demonstrate what electrification had to offer.

The co-operation continued and, as the rural networks grew, it became a practical proposition to extend mains electricity for the event rather than rely on mobile generators. The availability of mains electricity enabled the development of many side attractions and exhibitions, which, when added to the ploughing competitions themselves, helped to make for an enjoyable day out for all the family. It also provided lighting of the exhibition site thus eliminating many of the security and other difficulties previously experienced as darkness fell.

REO felt that such a popular rural occasion warranted something special in the way of promotions. The concept of a competition which would provide an attraction for women was mooted and so grew the Annual National Wholemeal Breadmaking Championship Finals which were to form an important side event to the ploughing competitions. There was a widespread belief that for the best home-made wholemeal bread a pot-oven was necessary. REO hoped to demonstrate that an electric oven could produce wholemeal bread as good and as tasty as the best from the pot-oven, with far less trouble and in a much more consistent manner. It was, therefore, a condition of the competition that all entries should be baked in an electric oven.

Preliminary competitions on a county basis were first run off by the various ESB Districts. The county champions, as well as receiving their county prize, had their expenses paid to attend the national finals. A large marquee was erected near the ploughing area and electricity laid on. A row of identical electric cookers was set up to bake the entries which were then evaluated by a panel of judges. The excitement heightened in the late stages of the competition as a packed marquee eagerly awaited the announcement of the winner.

There were some difficulties in holding such a competition in winter in the fields. One of the most abiding memories of REO staff is of constant rain, frequent gales, at least one snow-blizzard and almost always deep, sticky mud, as the fields on which the crowds circulated were almost as badly ploughed up by tens of thousands of gumbooted feet as were the ploughing competition areas. Sometimes it was a major effort to reach the bread-making competition marquee, isolated as it was in a sea of mud.

Such was the experience in the case of the first National Finals, which were held in connection with the NPA National Championships at Tramore in 1958. On the day before the event the newly erected marquee was torn to shreds by a ferocious gale and it took a tremendous effort by the local ESB staff to restore it to something approaching serviceable condition in time for the competition. There was a rather ambitious concept of actually grinding wheat in a small electric home grinder and using the resulting whole flour in the competition. Objections were raised by many competitors, as they were not used to this type of whole-flour and the idea was not repeated. The National Champion who emerged was Mrs Elizabeth Gorey of Burnchurch, Co. Kilkenny.

The following year there were two classes of competitors, senior and student. This resulted in a huge increase in entries, especially from students of technical schools. The senior winner of the finals, which were held in Burnchurch, was Miss Mary Turner of Carrick-on-Shannon, Co. Leitrim, while the champion student was 16-year-old Mary Kilgannon of Gleneaskey, Co. Sligo. A feature of the competition that was to continue was the involvement of the electrical trade and the farming press in the provision of prizes. The judges were Mrs M. Curtis, ICA, Miss E. M. Bonfil, Department of Education and Mona Fitzpatrick, Chief Demonstrator, ESB. These three ladies were to continue as judges in most of the ensuing competitions, a tribute to the integrity and dedication they brought to their task.

The senior competition was to continue without a break for over a decade – one of the longest-running joint promotions in which REO was involved – with venues at Oakpark, Killarney, Donea, Athenry, Danesfort, Enniskerry (which was mounted in a howling blizzard), Wellington Bridge, Tullow, Mallow, and the final competition was at Rockwell College in 1969. The student or junior section was run as an adjunct to the senior competition up to 1963, when the concept of a separately run schools competition was developed. This had the enthusiastic approval of the Department of Education, which arranged for the preliminary competitions to take place in the various technical and secondary schools. The National Finals of this competition were mounted on a dramatic scale under the joint banner of the ESB and the electrical manufacturing firm of GEC Ireland, in the Gaiety Theatre, Dublin, City Theatre, Limerick, Cork Opera House and in the new Abbey Theatre, Dublin. This competition was terminated in 1970 after a very successful six-year run.

CHAPTER SIXTEEN

The Voluntary Organisations

THE TASK OF REO in extending electricity and developing its use would have been much more difficult and would have proceeded much more slowly without the sustained help and co-operation received from the voluntary rural organisations. Four main bodies deserve special mention: Muintir na Tíre, Macra na Feirme, Macra na Tuaithe and the Irish Countrywomen's Association. The story of rural electrification would be incomplete without some account of these organisations and in particular the close co-operation over the years between them and REO in achieving what after all was a common objective. In a happy 'chicken and egg' situation the coming of electricity made it possible to develop attractive community meeting places with heating, lighting and facilities for drama, cine-projection, public address and similar aids, which in turn increased the effectiveness of the organisations and helped their growth and their contribution to the rural electrification effort.

MUINTIR NA TÍRE

The name of Muintir na Tíre (People of the Land) will always be coupled with that of its founder Rev. John Hayes, who launched this organisation for community development on a parish basis in May 1931. The growth and progress of the movement have been charted by Rev. Jerome Toner, OP, in his book *Rural Ireland, Some of its Problems*[1] and by Stephen Rynne in his biography of Father Hayes.[2] Father Hayes was committed to raising the whole standard of rural life. In a broadcast on 25 July 1945, he said: 'We realised that the land problem was not merely economic; the merely economic solves no problem. It was more than economic. It was the whole life of the people that needed stimulus.' It was a happy coincidence that the man appointed to head the Rural Electrification Organisation, W. F. Roe, had since 1938 been an active and prominent member of Muintir na Tíre dedicated to the ideals of its founder.

Both men saw in rural electrification one of the most effective means of providing the stimulus required to overcome rural stagnation. Speaking on 24 May 1948 at the switching on of electricity for the first time in Bansha, Co. Tipperary, of which he was then parish priest, Fr Hayes made this point. Having referred to the more obvious advantages of rural electrification he went on:

> . . . but there is a still greater advantage and that is the social aspect. No material instrument can do so much to uplift the rural people and give them a status in accordance with the important position they hold in the nation. Rural people who supply the fundamental necessities for the whole nation should at least have an equal right to the amenities of the nation. . . . It is more than an amenity, it is a revolution which will sweep away inferiority complexes.[3]

From the very start there was a very high degree of co-operation between Muintir and REO. The parish guilds were of the utmost assistance in initiating and organising the preliminary canvass to establish the priority for the official ESB canvass. In this the guild members made strenuous efforts to sell the advantages of electrification to often reluctant householders so as to achieve a high sign-up and a consequently high ranking on the list for official canvass and selection. In his book Fr Toner adverts to these efforts. Under the chapter heading 'Muintir na Tíre 1944-1949' he writes:

> The reports for these years show clearly that the war is over. There is a significant remark of 'Contacted REO', which means that the guild approached the Rural Electrification Office for the purpose of having its area considered under the terms of this Scheme . . . . The work of Muintir na Tíre in disposing farmers to avail themselves of this Scheme is of great importance. In the lectures and free debates arranged by parish guilds, much of the prejudice against innovations is broken down and farmers are encouraged by the confidence of others in the Scheme. Indeed, Muintir na Tíre would find it hard to develop rural Ireland without electricity. If the Scheme is a success, and that seems probable, it will bring about a complete change in Irish rural life. It will raise the general standard of living in country homes; it will speed up and increase agricultural production; it will induce capital to establish rural industries and it will brighten the social life of the parish. One may hope, therefore, that it will go far towards solving the problem of emigration; people leave rural Ireland because the land is not supporting its own. 'Contacted REO' then in a parish report . . . represents a fundamental work in the programme of the guilds.

The 'Rural Weeks' □ In January 1947 W. F. Roe received a letter from his friend Fr Hayes inviting him to give an illustrated talk on rural electrification and mount an exhibition of electrical appliances at a Muintir na Tíre 'Rural Week' in St McCartan's Diocesan College, Monaghan, in the following summer. (The Rural Week – more exactly a three-day rural week-end – was Muintir's big annual event and had been inaugurated in 1934 at St Joseph's College Roscrea.) Thereafter REO participated in many Rural Weeks where it sought to get the message of rural electrification across. In the following year at Carlow the main REO theme was on how to go about getting one's parish listed for consideration under the scheme. In 1952 at Summerhill College, Sligo, there were demonstrations in electric cooking. A film 'Running Water' was also shown each day to promote the concept of water on tap in the house and farmyard. In 1956 at the Killarney Rural Week the emphasis of the REO exhibition was on power tools suitable for farmers and for small community projects which might be set up by the local guild of Muintir. The main power-tool manufacturers co-operated to mount a comprehensive exhibition and working demonstration of equipment which included drilling, sawing, welding, brazing, ceramics and small stained-glass work.

This close co-operation in the Muintir na Tíre Rural Weeks continued until the abandonment of the concept in 1970 in favour of an annual national conference. It had been of mutual benefit, providing as it did an extra attraction in the Rural Week programme for Muintir and providing also a most valuable shop window to REO in which to display its wares and promote its message.

MACRA NA FEIRME AND MACRA NA TUAITHE

The war of 1939-45 highlighted the importance of the agricultural industry in Ireland and the pressing need to move away from the Cinderella image it had developed. In particular it was vital that proper agricultural education should be provided for those young people destined to work the land. The usual practice was that post-primary and higher education was reserved for the sons and daughters of farmers who would move out into other occupations. The son inheriting the farm usually received no formal education beyond that of the primary school. The small number of agricultural advisers and rural science teachers then existing were well aware of the necessity to provide a proper educational service in this area and did all they could in the face of limited resources, holding winter agricultural classes and rural science classes in the vocational schools.

In order to preserve the contact set up between pupils and teachers at these winter classes, the idea of young farmers' clubs was born and in

1942 and 1943 the earliest such clubs, known as farmers discussion groups, were formed. These led in 1944 to the formation of a national organisation of young farmers' clubs, with a central executive. In September 1947 a national headquarters was opened by Seán T. O'Kelly, President of Ireland, in Athy, and in December the title Macra na Feirme was officially adopted. In 1948 a fortnightly paper entitled *Young Farmers Journal* was launched, which was destined to be the forerunner of the very successful *Irish Farmers Journal*.

At this period the question of an organisation to speak for all farmers on economic issues was raised. It was decided Macra na Feirme would concentrate on the educational, cultural and social interests only and in 1955 a separate organisation, the National Farmers' Association, later to be re-named the Irish Farmers' Association, was founded to look after the economic interests of farmers.

From the very start Macra na Feirme gave very strong support to the concept of rural electrification. Like Muintir na Tíre, it helped to organise local canvasses and to support REO in the official canvass and selection process. There was a great deal of dual membership and in many cases the two organisations worked closely together to try to ensure early selection of their parish or area. On the odd occasion a degree of rivalry broke out, as when a contest developed in 1948 between the parish of Cahir and that of Bansha as to which would be the first area to be developed in County Tipperary. The prime mover in Bansha was Muintir na Tíre, led by its founder and parish priest of Bansha, Fr John Hayes. In Cahir, Macra was the main organiser and when Bansha was selected on the basis of a better economic return, although Cahir had a better sign-up, a deputation from Cahir called on W. F. Roe to protest against the alleged favouritism.

Despite the occasional controversy, relations between Macra na Feirme and REO continued on a most co-operative basis. REO staff attended with films and demonstrations at the various Macra functions, gave lectures on the applications of electricity at their winter programmes and co-operated in organising competitions, while Macra members in their turn strongly supported REO's efforts to extend and develop the Scheme. On at least one occasion the REO stand at the RDS Spring Show was designed and staffed by members of Macra.

In 1951 the idea of a junior Macra to cater for the twelve to eighteen age group was born. The main thrust came from a number of rural science teachers in vocational schools, but it immediately received strong backing not only from Macra na Feirme, which gave it particular organisational support in its early days, but also from Muintir na Tíre and the Irish Countrywomen's Association. The new organisation was given the title of Macra na Tuaithe. It was broadly based on the same concept as the American 4H Clubs with strong emphasis on the principle of learning by doing. The members of the new organisation were the young people who were to remain on the land. They were bright, forward-looking and eager to experiment with new methods. The promotion of Macra na Tuaithe throughout the country was helped immensely in 1958 by a grant of £30,000 from the American Kellog Foundation, to be used for this purpose over a five-year period. In order to continue the financing of the organisation when the Kellog Foundation grant ran out in 1963, the National Youth Foundation was established, to which Irish industries and commercial concerns were invited to contribute an annual sum. Over 200 firms responded positively. The ESB became a member of the Foundation from its inception and also took part in a scheme of national awards, which had been set up to encourage initiative and effort among its members. In addition to its membership of the Foundation, the ESB also agreed to sponsor an annual Home Improvement Award for Macra members. This award was for the planning and carrying out of an interior decoration scheme, and the ESB was joined in its sponsorship by the firm of Wallpapers Ltd. In the same year the ESB also agreed to contribute towards the costs of an annual Macra Leadership Course, involving fifty participants, at the ICA residential college at An Grianán. Inspired by the enthusiasm of its founders and supported by both its industrial sponsors and the State, through the Department of Education, Macra na Tuaithe flourished. In 1959 there were 125 clubs, in 1963, 231 clubs, and by 1966 there were 253 clubs dispersed throughout the whole twenty-six counties.

The Citizenship Award □ In 1967 the Home Improvement Award was discontinued owing to difficulties associated with its supervision and in 1968 a new competition was sponsored by the ESB entitled Know Your Area. As originally proposed, this was a club project with four prizes of £50 each for the winning club in each of the four provinces, the four provincial winners competing for the national prize of £100 at the annual summer gathering. In 1969 the *Irish Farmers Journal* joined in the project providing additional welcome finance and it was decided to award, in addition to the cash prize, a perpetual trophy and replica for retention by the winners. For this revised competition now known as the Know Your Area Citizenship Award, the provincial basis was revised. County Donegal was included in the Connacht group and Cavan and Monaghan were included with Leinster, so giving three geographical divisions. The perpetual trophy took the form of a carving in cherrywood and bog oak.[4]

The Citizenship Award competition was based on Macra's Citizenship Programme and involved the collection, organisation and recording of data on one or more aspects of life in each club's area. Director Michael Cleary speaking in 1970 expressed the purpose of the Citizenship Programme as 'to help young people acquire the skills necessary to maintain and develop a true democratic society. Democracy implies that people must be informed citizens. Our project trains young people in the methodology of studying their society and, even more important, in doing something about the problems in their society once these have been identified.'

The Programme and Award developed a whole new concept as to what the organisation should do. It had originally been founded to promote, principally at least, rural science and home economics as extra-curricular activities from the local vocational school. Arising out of the Citizenship Programme came a new concept: a more active outgoing role in local society. From knowing one's area came the transition to action. In the early stages of the competition the main emphasis was on investigation of the local scene and possibly criticism of what the investigation turned up. The more active clubs were not, however, content with just this; they initiated action programmes to correct or improve the situation as they had found it. In later years it was an essential part of the competition that the club should become involved in some type of action.

From 1971 on, the Award was sponsored by the ESB alone. It unearthed a formidable array of talent among the youth of rural Ireland. From the very beginning, the standard of entry was high and the standard of the winners and runners-up approached that of professionals. The winner of the Award in 1969, its first year, was the Ballyfin, Co. Laois, club. The runners-up were Grange, Co Sligo and New Inn, Co Tipperary. The sponsorship of the Award by the ESB continued even after the official termination of the Rural Electrification Scheme and the Citizenship Programme still plays a central part in the life of the clubs. The variety of projects is endless: local effects of pollution, history of parish, relationship between nutrition and health, local implications of EEC membership, formation of mussel-farming co-operative, tidying up of local cemetery, study of old people's problems, promotion of active interest in sport.

In 1981 the title of the organisation was changed from Macra na Tuaithe to Foróige or the National Youth Foundation, to reflect the fact that it was no longer an exclusively rural body. Clubs had now been set up in urban locations and it is interesting to note that in that year the Award was won by an urban club, The Cool Kids in Ballymun, Dublin, in its third year of existence. The project, Operation Access, was chosen in the Year of the Disabled to help the public realise some of the difficulties facing disabled people. Having acquired a number of wheelchairs, members invited both shoppers and shopkeepers in the local shopping centre to sit in the wheelchairs and try to do some of the things normally taken for granted such as shopping and making telephone calls. The findings shattered the complacency of the community and brought about a greater understanding of the problems of the disabled. Improvements were demanded in many areas. An immediate result was that the local supermarket widened the space at the check-outs to allow wheelchairs to pass through.

The finals of the competition and the presentation of the Award were always the highlight of the year's activities and took place at the organisation's summer gathering, usually at Gormanston College. An indication of its prestige was the fact that on various occasions the Award was presented to the winning team by the President of Ireland, Erskine Childers, and the two religious primates, Cardinal Conway and Archbishop Simms.

No one project so vividly illustrates the social and educational progress of rural Ireland in one generation as the Citizenship Programme. From the stagnation of the late 1940s to the dynamism of the youth of the late 1970s was a tremendous leap in which the provision of amenities made possible by rural electrification played a vital part. Without these, the task of the voluntary bodies such as Foróige would have been extremely difficult, if not impossible.

THE IRISH COUNTRYWOMEN'S ASSOCIATION

Sir Horace Curzon Plunkett, father of the co-operative movement in Ireland, writing of his initial difficulties admitted to making a couple of fundamental mistakes:

> But we [the founders] failed to realise fully two things: the first was the enormous difficulty of establishing a new social organisation in a country where there is none already existing; it is like digging foundations in sand. The second was the importance of women's work. We had put 'better business' first, 'better farming' next, and 'better living' was to follow as a result of these two, but, as things are to-day, we shall not get very far either with better business or better farming unless we can show some better living as a bait and thus stimulate the desire for more. And this the women can do best; in fact they alone can do it.[5]

Plunkett's experience was not lost on the strategists of REO. The ability and motivation of the women to achieve the 'better living' referred to by Plunkett and to spur on the often reluctant male partner to accept and utilise the new electricity for this purpose was manifest. In developing co-operation with the Irish Countrywomen's Association, REO was pushing at an open door: it was precisely to achieve this raising of the standard of living in rural Ireland that the ICA was founded in 1910 under the name of the Organisation of United Irishwomen. The organisation had its roots in Plunkett's great co-operative movement which had up to then been male-dominated, but, to quote Ellice Pilkington, one of the founders of the United Irishwomen:

> I was recommended by a relative of my own, a member of the Irish Agricultural Organisation Society to attend the Annual General Meeting of that Society in 1909 .... Many women who are now 'United Irishwomen', but who were then entirely unconnected with the Agricultural Organisation Society, were also present. Mr George Russell (AE) at that meeting read a paper on rural civilisation and in it pointed out to us quite clearly that our place was ready for us if we were ready to step into it.[6]

On 15 June 1910, Mrs Harold Lett, Vice-President of Wexford Farmers' Union formed the first branch of the United Irishwomen at Bree, Co. Wexford, a non-sectarian, non-political organisation affiliated to the IAOS. This was followed by a visit to Paddy the Cope in Dungloe, Co. Donegal, in December, resulting in the formation of a branch there of two hundred members. Four more branches followed in Wexford and then one in Kilkee, Co. Clare. The Society of United Irishwomen had been successfully launched.[7]

It was stressed from the beginning that the Society was strictly non-political and non-sectarian and intended to include all levels of society. Nevertheless, and inevitably, the title United Irishwomen appeared to many to imply a political connection and it was decided to change it to the Irish Countrywomen's Association, or in Irish Bantracht na Tuaithe. By 1969 there were 22,000 members in 877 local guilds throughout the country.

The purpose of the new organisation was to organise women in every district in Ireland for community service and to improve the country's domestic and reconstruct its social life. To quote Plunkett once more:

> ... they come in to complete our work where we believe it to be at once most important and most incomplete . . . and as one of those responsible for this imperfect movement, I hope they may realize the most cherished and least fulfilled ambitions of its male leaders, the evolution of a healthy

> and progressive community life. To women's influence I look for the brightening of the social sky in the grey dawn of peasant Ireland.[8]

Dr Muriel Gahan in a lecture given in 1969 takes up the theme:

> 'The evolution of a healthy and progressive community life' — what a task in the year 1910. No wonder the men had despaired of their better farming and better business alone bringing it about. Nothing but a woman's dauntless spirit, a cause to which she was utterly committed, and a programme dealing with needs as she met them, could have made any of those pioneer countrywomen confront the rural problems of the day. With prejudice of every kind against them, these women did confront them under their chosen banner 'Deeds, not Words' . . . Countrywomen's present day needs, with which among others, our Association concerns itself, for more piped water and labour saving houses, home advisers, adult education and training, have developed from those basic needs met by the first United Irishwomen, District Nurses, adequate food and clothing, with milk as a priority.[9]

The advent of rural electrification brought nearer the prospect of piped water and labour saving houses to Irish countrywomen and the Association threw itself into a full scale campaign to promote these. Piped water was regarded as the top priority, as with it would come the end of the drudgery of hauling buckets of water from the well, a task too often relegated to the rural housewife. With it also would come a transformation of the sanitary facilities which even in the late 1940s were, in the case of nineteen rural homes in every twenty, of an extremely basic nature.[10]

Thus it was that at the ICA Annual Fair at the Mansion House Dublin on 8 and 9 November 1950, a special display, with water as the central theme was mounted by the ESB Rural Electrification Office. The centre-piece, designed to make a strong initial impact, was a huge aluminium water tap out of which water was gushing at the rate of 2,500 gallons per hour. To supplement this was shown a full size replica of a corner of a cottage kitchen with sink and hot and cold water. James Dillon, Minister for Agriculture, was photographed in the act of turning on the huge tap. Later he spoke of the importance of running water and pointed out the assistance available from his Department. His concluding remarks were highly applauded: 'Young ladies of the country; make it known that there will be no more marriages until there is hot and cold water on tap in the kitchen'.

Dr Muriel Gahan, champion of the spread of rural electrification, who was dedicated to the improvement of the lot of the Irish country woman. She is holding an AIB community service award, 1974.

The ESB/ICA Kitchen □ Apart from water, the importance of good design and layout in the achievement of a labour-saving home was continually stressed by the ICA. In the kitchen particularly, where the rural housewife spent so much of her time and did so much of her work, a well-designed layout could do much to eliminate traditional drudgery. In 1956 the Minister for Agriculture, again James Dillon, requested the ICA to set up a typical farm kitchen of the future as part of the Department's exhibit at the RDS Spring Show. Once again the ESB joined in and the ICA Farm Kitchen was born. It showed how a typical farmhouse kitchen could be renovated to become a comfortable living and working centre for the whole family. The following year, as the opening of the Simmonscourt extension made more space available for the ESB exhibit, again the ESB and ICA, with architect Eleanor Butler as consultant, joined in designing and furnishing a modern labour-saving farm kitchen. Such was the interest shown that it was decided that the kitchen should be made available for exhibition throughout the country, and in March 1958 the ESB/ICA Mobile Farm Kitchen took to the road.

An itinerary was devised so that the kitchen could be exhibited at strategic locations throughout the country. It was open to the public for a certain period each day, but at night was reserved for members of the local guilds of the ICA, Muintir na Tíre and Macra na Feirme, who received lectures from kitchen planning experts of the ICA and from technical instructors who showed how the kitchen was a conversion of a traditional rural kitchen and how it could be built up over a period, long or short, on the do-it yourself principle. There were also demonstrations of electric cooking and general kitchen craft. The exhibit was built on a special trailer chassis, which had to conform to road traffic requirements when in transit. On reaching its selected site, however, it opened out to form a spacious kitchen which provided for a food preparation and serving centre, ample cupboard space, sink area, drop-out ironing board, and, in addition to an electric cooker, a hearth to retain the traditional atmosphere of the old fashioned kitchen. Off the kitchen proper was a bathroom and laundry area.

For many years the mobile kitchen worked its way up and down the country being seen and discussed by young and old alike. Special essay competitions were organised for school children on what aspects of the kitchen impressed them and small prizes awarded. Sometimes it provided a community service as when on a cold night in December 1958 in Dunfanaghy, Co. Donegal, sandwiches and tea were prepared and served in the kitchen to 120 participants by a ladies committee on the occasion of a 'twenty-five' card drive in the local hall in aid of the school building fund. Some weeks later when the kitchen was making its last stop in connection with the ICA winter programme in County Donegal, it again came to the rescue in Derrybeg, when the water supply to the local hotel failed, knocking out the solid fuel water-heating and cooking system on a day when a large funeral had resulted in a heavy demand for luncheons. With the help of the mobile kitchen parked nearby, the ESB man in charge, Colm Browne, arranged immediate alternative cooking facilities for the hotel staff ensuring an excellent luncheon to all of the customers and earning the plaudits of the local community.

The kitchen did not as a rule attend special functions such as shows where it would be only required for one day, but on the occasion of the Annual General Meeting of the Wexford Federation of the ICA, held in Coolattin on 11 June 1959, and coinciding with the Coolattin Fete, an exception was made. The opening ceremony was performed by 'Biddy Mulligan' (the late actor Jimmy O'Dea), who was met by Olive Countess Fitzwilliam, on whose estate the Fete was held, at the door of the kitchen. Over 2,000 visitors passed through during the remainder of the day.

In these and in many other ways the mobile kitchen was kept to the fore in the media ensuring that wherever it went it drew large crowds. On the fairday in Tullamore in May 1959 a farmer called in to the kitchen which was on exhibition and offered £1,000 in cash for it as it stood! There is little doubt that the subsequent raising of the standard of Irish rural kitchens as pleasant and efficient places in which to live and work was due in great part to the ESB/ICA kitchen in its various forms, mobile and static, reinforcing as it did in a most effective way the continuing educational efforts of the Association.

The An Grianán Scholarships □ In 1953 the Kellog Foundation of America made a gift to the ICA of a large country mansion in Termonfeckin, Co. Louth, for use as a residential college. Apart from the main house there was a small cottage adjoining the farmyard and the idea grew that this could be refurbished and made to resemble a modernised farmhouse, incorporating the design of the mobile kitchen as far as possible, and used as a training unit for young rural housewives of the future. The ESB agreed to provide the electrical equipment and to award fifty-four scholarships annually from 1954 to cover six-week courses for nine girls at a time, while the ICA provided the training. The candidates were from the vocational school courses in the various counties.[11]

The scholarships were confined to girls who had been successful in their courses at the vocational or techincal school and who were most likely to return to their homes after the course, having acquired new skills as homemakers. Selection was carried out by the Department of

Education. Emphasis was on the practical use of electrical aids and the course covered cooking and household accounts (with special emphasis on budgeting), laundry and home management, and poultry, dairy and gardening.

A further set of twenty-seven annual scholarships was sponsored by the ESB for senior ICA members, again in homemaking, with emphasis on the use of electrical aids to relieve drudgery and improve the standard of household management. In the farmyard a building was reconstructed and converted into a cafeteria and folk museum with a working replica of the ESB/ICA Farm Kitchen as the focus. Here also the ESB sponsored specialised courses such as the 'farm guesthouse' course for farm housewives wishing to join the Bord Fáilte scheme.

The ICA was very interested in the concept of promoting wholemeal bread baked from home-milled wheat as an economical and nourishing item in the diet of the rural community, and a feature of many demonstrations with the Farm Kitchen was the production of genuine stone-ground wholemeal flour ground in the 'Barngold' home-grinding mill and the baking of the wholemeal bread in the electric oven at a total cost of about 2*d.* per pound of finished bread.

In May 1959 a 'Rural Family Week' was organised by the ICA in Rathmines Town Hall in co-operation with the Home Economics Section of the Food & Agricultural Organisation (FAO) at which the role of electricity was given prominence. A paper entitled 'Electricity and the Rural Family' was read by P. J. Dowling, Engineer-in-Charge, REO.

While the ESB, the ICA and the Departments of Agriculture and Local Government were each working to promote the extension of piped water supply to rural homes, progress was painfully slow in view of the huge numbers still to be served. The ICA decided that a much more intense and unified campaign was necessary if worthwhile progress was to be achieved. It therefore called a meeting of all interested parties on 24 August 1960 at its headquarters in St Stephen's Green to launch a national campaign to develop rural water supplies. The ICA notes in the *Farmers Gazette* of 3 September 1960 announcing the campaign stress that 'every Irish countrywoman must become a preacher in this campaign. If each and every one of us lifts her voice and demands water for rural homes, the present scandal will be remedied, for it is a scandal that only 12 per cent of rural homes, have a piped water supply.' (See chapter 17 for a more detailed description of the campaign.)

The Drimoleague Kitchen □ The Area Organiser for the Cork County Federation of the ICA, Mrs Kathleen Gleeson, had seen in Holland the work of teaching rural housewives new and better ways of housekeeping carried out by a similar local organisation in the kitchen of a house belonging to a selected resident of the area. In the summer of 1959 the idea was adopted in Drimoleague, Co. Cork, in the house of Mr and Mrs Humphrey O'Leary, two years married, who were in the course of remodelling their kitchen. The ESB agreed to sponsor the project and supply the electrical equipment. Firms in Cork also gave asssistance by providing tiles, paint, pots and pans etc. while the Cork Vocational Education Committee agreed to provide a weekly class in domestic economy. A committee representative of the various groups co-operating was set up to run the project. Sixty local women applied for enrolment for classes held in the environment of a well-designed and well-equipped kitchen. The project operated succesfully for the agreed period of two years after which it was terminated to prevent its becoming a burden on the very co-operative owners. At the winding up ceremony the chairman of the operating committee was enthusiastic about its value: 'The example of kitchen planning has been followed in the houses of the district while its good influences on the social side should not be omitted — the local men and women met and discussed their problems having a cup of tea between lectures in the Electric Farm Kitchen.'

In 1961, a similar project was launched at Castlemaine, County Kerry, in the home of a member of the local ICA Guild, Mrs Timothy O'Brien. Classes in domestic economy were held twice a week from 1 March to the end of June with the co-operation of the County Kerry Vocational Education Committee. An average of about forty, mostly ICA members, attended. In 1963 members of Macra na Feirme and Macra na Tuaithe were involved as the ESB commenced a series of weekly demonstrations in the use of electricity in and around the home. The afternoons were for Macra na Tuaithe (twenty-nine boys and six girls attended), while the evenings were for the seniors with an average attendance of fifteen men. After the agreed period of two years, the project was terminated.

By the completion of the Rural Electrification Scheme in the mid-seventies the ESB/ICA partnership could look back with satisfaction on the progress achieved towards 'better living' for the rural community. Electricity was available in almost every home, the standard of rural housing and housekeeping had improved and water on tap was well on the way as a reality for all. However, the task is not yet finished and while television has largely superseded the meetings and gatherings in the local hall as a source of entertainment and information, the close co-operation still continues in many areas particularly in the sponsorship of courses in An Grianán.

CHAPTER SEVENTEEN

Water on Tap

Part of the drudgery of the woman's work on the farm was attributed to the farmhouses. The long low thatched building, with the well a field or two away, and the candle in the window at nightfall — all that might be something for homesick Yankees to enthuse over, but it was a heartbreak for the unfortunate woman who had to keep it clean: who had to cook on its back-breaking hearth, while watching her children blinding their eyes trying to get their lessons done by candle or oil lamp light: and it was she who had to tramp the muddy path to the well, and carry back every bucket of water needed for her work.[1]

EVEN THE MOST ARDENT ADVOCATE of electrification on the REO staff would have to admit that running water should take priority in a rural home over the other benefits of electrification; but of course it was the coming of electricity that made it possible for the first time to install water on tap in the majority of rural dwellings at a reasonable cost. Electricity is but a means to an end, the end in this case being the raising of living standards, the elimination of drudgery in the farmhouse, the farmyard, and on the farm, and the increasing of productivity and, consequently, farm income. What better first step towards these objectives than the installation of an electric pump.

Running water could do much to raise living standards. A footnote in the Report of the Commission on Emigration and Other Population Problems refers to the absence of water on tap or of proper sanitation in the Rural Ireland of the late 1940s:

In the 1946 Census Report on the matter it was stated that in rural areas in the twenty-six counties only 9 per cent of households have a piped water supply, 44 per cent get their water from private wells, 26 per cent from pumps, 12 per cent from streams and the remaining 9 per cent from other sources. Half of the farm dwellings get their water from private wells but only 5 per cent have a piped supply laid on to the buildings and these are all on farms over £200 Poor Law Valuation. Only a little more than half

> the number of farm dwellings with a piped water supply laid on have a fixed bath. As regards sanitary facilities, four out of every five farm dwellings in the twenty six counties have no special facilities such as flush lavatory, chemical closet, privy or dry closet, and of the very large dwellings, (P.L.V. £200 and over), about one seventh have no special facilities. Of the small number with facilities (one out of every five), only about a quarter have indoor lavatories. Thus, only one out of every 20 farm dwellings has an indoor lavatory.[2]

Drudgery was an inevitable outcome of the situation. In over 90% of homes every drop of water used in the house had to be carried over the threshold, most of it from a considerable distance, usually by the farmer's wife or children. Every drop of hot water had to be heated over the fire. Most Irish farm housewives set high standards of cleanliness, but under such circumstances the achievement of this involved an inordinate amount of time and effort. To quote from John Healy: 'on churning day she'd [Grandma] scald everything with iron kettles of hot water. I would run to the well several times that day and the last can of water went for the washing up afterwards, when everything had been scalded again.'[2] When account is taken of the time spent on the multitude of other chores around the farmyard, traditionally the responsibility of the woman of the house, it will be seen that the lot of the average farmer's wife was a hard one. It is no cause for wonder that so many farmers' daughters opted for urban life and urban husbands, rather than marry farmers and inherit the life of unremitting drudgery that had been the lot of their mothers.

Just as running water in the house would emancipate the farmer's wife from so much drudgery, so out in the farmyard and in the fields it could do the same for the farmer. For the dairy farmer in particular it meant a higher standard of hygiene and subsequently higher prices for his milk. For his cows a continuous supply of drinking water could mean higher milk yields and for dry stock, better weight gain, not to speak of the time saved in watering stock on so many farms. It was natural then that from the very outset REO put the greatest emphasis on the provision of running water on the farm as one of the first fruits of electrification. At every lecture, demonstration or other promotion of the Rural Electrification Scheme water pumps and water systems were given pride of place. Very quickly a pump advisory service was set up, which developed into a comprehensive water advisory service which gave advice and assistance on all aspects of running water installation from the location of water sources to the selection, installation, and maintenance of pumps and water systems. This advice and assistance was made freely available, not only to individual householders, but to group organisations, local authorities and government Departments. Thirty-five years after its foundation the service, now part of the Agricultural Advisory Unit of the ESB, is still contributing to the national drive for the installation of running water in every home in the State.

REO was conscious that preaching the advantages of electric water pumping was futile unless the actual appliances were made easily available and co-ordinated efforts made to install them in consumers' premises. From the outset, therefore, even in the immediate post-war years when supplies were scarce, a policy of stocking pumps of a size suitable to the requirements of rural consumers was followed. The emphasis was on reliability and simplicity of installation, plain equipment with no frills. At the rural electrification stand at the Spring Show of 1948, three systems for obtaining water at the kitchen sink were shown. All could be installed by the average handyman. The first and most basic exhibit was designed to take the mystery out of installing a water supply. It consisted of a small centrifugal pump mounted over a 'well' and driven by a belt from a fractional hp motor. The output from the pump was brought in over the kitchen sink by means of a ½" pipe and discharged directly from the pipe into the sink. Not even a tap was involved. A cord switch over the sink switched the motor and pump on and off as water was required. The ultimate objective was, however, to have hot and cold water on tap and the second example showed how a self-contained electric pump could feed a roof tank being switched on and off automatically by means of a float switch as the tank filled and emptied. This could form the basis for a combined hot and cold water supply by adding an electric water heater. The third exhibit was a pressure-storage system with a tank of 35-gallon capacity which was completely self contained, operating at a pressure of 20 – 40 lbs per square inch, which required no roof tank and indeed required the minimum of structural work to install. This type of system was to prove by far the most popular. Over the years tens of thousands were installed using different types and sizes of pumping and storage units to meet the customers' requirements.

At the 1950 Spring Show one of the REO stands was given over completely to water supply and full publicity was given to the Department of Agriculture grants that were available. Both shallow-well and deep-well pumping systems were shown actually working. In addition, two film strips were shown continuously, one on How to Bring Water Supply to the Farm and the second on Hot Water on the Farm.

As the Rural Electrification Scheme progressed no opportunity was missed to promote the message of running water in the home and on the farm. Members of the Irish Countrywomen's Association were very

Above: Carrying water home — ESB advice below is 'let it run'. *Below*: James Dillon, Minister for Agriculture, turning on the giant REO tap at the ICA Annual Fair, Mansion House, Dublin, November, 1950.

powerful allies. Time and time again pressure from the woman of the house was the deciding factor in persuading the man of the house to invest in a water system. Once the first steps had been taken the rest followed quickly. Starting with a sink and cold water tap in the kitchen a hot water installation soon followed. A flush lavatory was then installed and a bathroom. Supply was quickly extended to taps in the farmyard and, in places where streams were scarce or unreliable, out to the fields to ensure a constant and reliable supply of drinking water for the stock.

The importance of efficient servicing (diagnosis of problems, repair of faults etc.) was recognised at an early stage and training courses for ESB electricians in maintenance and repair of pumps were instituted. A comprehensive series of articles on pump trouble-shooting was published in *REO News* and specialists in all aspects of water pumps made regular visits to local centres.

In pre-war years resources for the provision of public piped water supplies were of necessity confined almost exclusively to urban areas. The emphasis was on the improvement in public health conditions and priority lists were drawn up, which naturally concentrated firstly on the more densely populated urban locations as presenting a greater potential public health hazard than the sparsely populated rural areas. Just before the 1939-45 war, however, a start was made in supply to rural areas on a regional basis in the Rosses, Co. Donegal, a thickly populated area with poor local water sources. The war put an end to such expansion and in the immediate post-war years when capital for public investment was scarce, roads, housing and electrification itself received priority over public water supply in rural areas. By 1950, however, there was again movement and a system of contributions from the central government to the cost of local authority water supply schemes was introduced. In the same year a system of grants for private installations on farms was initiated by the Department of Agriculture, followed in 1952 by a similar grant system for private houses by the Department of Local Government.

The Department of Agriculture scheme provided for a grant of 50% of the cost (up to a maximum grant of £100) of providing water suitable for domestic use and piping it to kitchen sink and water tap in farm dwellings. In initiating the scheme the Minister for Agriculture, James Dillon, appealed particularly to the wives and wives-to-be of farmers 'to employ the maximum diplomatic pressure, which connubial propriety, or the privilege of bethrothal may properly allow, to persuade the head of the household to provide this indispensable amenity of a piped water supply'. These grants paved the way for the installation of private schemes but, despite the forceful appeals of the Minister, progress was painfully slow in the early years. In the year 1950/51, only 150 water pumps and pumping systems were sold to rural consumers by the ESB. In 1953/54 the total sales for the year were about eight-hundred pumps (of which the ESB share was 350), a pitifully small figure in view of the strong promotional effort. A sub leader in the *Irish Independent* on 19 May 1954, reporting that in the third year of the grants being made available only eight hundred grants had been paid, gloomily calculates that 'unless this figure can be improved on, it will be nearly 500 years before every farm has water on tap in the kitchen'.

However, due to the combination of grants, the support of the voluntary rural organisations and intense promotion, particularly by REO, the rate of connection rose, albeit slowly. Sometimes it was helped along by a little judicious blackmail on the part of the woman of the house. In the Oldcastle area the Area Organiser's notes for April 1955 record that one farmer's wife in his area had informed her husband that she would carry no more water in buckets if he did not sink a well before the end of June. 1956 saw 1,100 pumps sold by the ESB and by 1959 a Local Government committee reported that 12% of rural dwellers now had piped water supplies, 9% from private sources and a further 3% served by public supply. The total still amounted to only about 50,000 houses, while at this stage electricity had been extended to almost 250,000 dwellings. To quote from an article in *The Manchester Guardian* on 28 April 1959 'It is still possible [in Ireland] to see a man balancing two buckets of water from the pump as he hastens home to be in time for some favourite television act. His house may have a vacuum cleaner, but not a tap.'[3]

Whatever the reason for the slow progress in 'rural aquafication', it could not be attributed to lack of effort by REO. In almost every issue of *REO News,* at every rural electrification demonstration, in REO newspaper advertisements, the message of water on tap was promulgated. When in 1955 the problem of locating suitable sources of water was identified as a major problem, arrangements were made between REO and the Geological Survey Office to provide a water location service. On any particular farm or in any particular locality the GSO would indicate on a six-inch map the most likely water sources and give an estimate of the depth at which water would be found and the amount available. This arrangement was of mutual benefit, as the borehole results on site were fed back to the Geological Survey thereby enabling them to update their records. A still further advance in the finding of water came with the advent of borehole-contractors prepared to operate on the basis of 'no water, no fee'.

Field demonstrations of the actual installation of a pumped water system on a farm were also part of the REO promotional effort. Attendances at these by local farmers were high – up to 100 in some cases – and here they could see for themselves how simply and comparatively cheaply a system could be installed.

In the years following the war suitable pumps and pumping systems were, like most equipment, difficult to secure. The most suitable unit for the average small farm was a shallow well pressure-storage system operating at 20 to 40 lbs per square inch, which in 1948 sold for £32. At first these were obtained from the Everite Company in the USA, but devaluation of sterling and the dollar shortage in 1949 caused REO to look elsewhere. After trying out a large number of alternatives, a German pressure-storage system using a piston pump, the Leöwe 'Wasserknecht' (later called 'Waterpak'), was selected and many thousands of these were installed throughout the country over the next couple of decades.

Shallow well pumps, however, depending on suction alone, would only lift water from a depth of about twenty-five feet and pumps suitable for deeper wells were also required in increasing quantities. One of the most versatile of these was the jet pump introduced by REO in 1951 as part of the American Everite range. This fed some of its output back down the well under pressure. At the bottom of the well a special arrangement utilised this fast-flowing jet of water to force fresh well water to the surface. In their early stages of development, jet pumps were able to pump from depths of around eighty feet. They were simple in construction having a minimum of moving parts. For extreme depths and very high output, submersible pumps, tailored to fit bores of down to four inches in diameter and which could be installed at the bottom of boreholes, were available. Depending on the size of the pump and motor, these were of almost unlimited capacity and could operate from great depths. Hence they were very suitable for supplying large schemes from deep boreholes.

In 1953 Irish-manufactured pumps became available when the firm of Unidare entered the domestic pump field with a unit based on the centrifugal jet principle. This was initially available as a shallow well or deep well automatic pressure-storage system with a maximum operating depth of eighty feet which met the requirements of most wells bored at the time. As the demand for bigger pumps, higher operating pressures, and deeper wells grew over the 1960s larger units were developed, culminating with capacities of up to 2,500 gallons per hour, operating pressures of up to 100 lbs per square inch and lifting from depths of almost four hundred feet.

PIPING

Another obstacle to the installation of running water in the post-war years was the scarcity of piping. Although this was somewhat outside the normal field of electrical merchandise, REO took steps to locate and stock supplies of steel tubing so that work on installations was not held up. The jointing of this tubing, which was available only in comparatively short lenths, was tedious and, since it involved the cutting of threads, required a certain degree of skill which was rather off-putting to the average householder. Two developments, however, helped to clear this bottleneck and contributed immensely to the spread of rural water supplies.

In 1953 the firm of Unidare in Finglas, Dublin, commenced production of their 'Hydrodare' plastic water pipe. In contrast to the steel piping, this was cheap, strong, light and corrosion-proof. It was reasonably flexible and could be supplied in coils of considerable length, which reduced the problem of jointing – in any case a comparatively simple process – and it lent itself to mechanised methods of laying. In 1958 the Dutch company of Wavin opened a small factory in Drumcondra, Dublin, to manufacture rigid PVC water piping in 20-foot lenghts. Again the piping was extremely light, strong and corrosion-proof; it could be easily jointed and was available in a wide variety of bores. In 1962 the company moved to a much larger premises in Balbriggan, Co. Dublin, and in 1971 commenced manufacture of flexible piping. The ready availability of these two Irish-made products removed one of the biggest obstacles to the widespread extension of piped water in rural Ireland.

THE GROUP SCHEMES

It was obvious from the start that the extension of piped water on the basis of an individual pump to each farm would be a slow process. In addition, in many parts of the country reliable sources of water were difficult and expensive to develop, prohibitively so if for only one user. While the government and the local authorities were dedicated to providing public water supplies on as extensive a scale as possible, their immediate priorities were the construction of new and the improvement of existing supplies in the expanding urban areas and in some of the more thickly populated rural communities. It would take many years before they could direct their resources towards the more sparsely populated country areas. This situation led to the development of the Group Water Scheme, under which a number of adjoining householders co-operated in developing a common water source.

The first record in REO of this kind of co-operation between neighbours was in 1956 in the Kilfinny Rural Area of County Limerick. James

Cronin, an engineer in the ESB, whose parents farmed in the area, suggested that his father and two neighbours should locate a reliable common source of water and install a common pump to supply water to their farms. Drinking water had to be carried a long distance. Water for farmyard use had to be carted during the summer droughts and in the winter, when the cows were in their 'pounds', the only supply available in the farmyards was rainwater collected in large concrete tanks. Each farmer had a few wells on the farm, but these invariably failed for three months each summer, fed as they were by surface water from nearby drains. Full co-operation was given by REO in selecting and providing a suitable pumping system and advising on the layout of the supply. A diviner was employed who located a suitable source and after some initial reluctance, an agreement was drawn up and signed by the three parties.

In 1957 a big breakthrough came at Manor Kilbride, in Co. Wicklow, where a young curate, Rev. Joseph Collins, was anxious to organise a communal water supply for thirteen families, including five farming families at Old Court where the only water available was from a county council tap half a mile from the centre of the group. To quote his own words:

> . . .somebody mentioned that the Department of Agriculture was giving grants for the installation of private water supplies and the idea was born that by combining a number of these grants it might be possible to extend the Co. Council pipeline down the road. The Department of Agriculture's Inspector was approached and agreed that while in theory the scheme was practical, in actual fact it was not, because there were only five farmers among the thirteen families.
> However, he referred us to the Housing Section of the Department of Local Government which gives grants to every householder – farmer or otherwise – for the installation of a piped private water supply. A telephone call to the Department of Local Government confirmed this and to the further question of whether it would be possible for a number of householders to have these grants and install a joint water supply came the cautious reply that it had never been done before, that prima facie there was nothing against it but that further enquiries would have to be made in the Department before a definite reply could be given. In due course came the official reply that such a scheme would have the blessing of the Department.[4]

Fr Collins also got agreement to a most important principle: contributions of labour to the scheme could be evaluated and included as costs for the purpose of the grant. The result was that by the end of August

1958 every one of the thirteen houses had a sink and piped water on tap in the kitchen. The cost per house was £17 plus the Local Government grant of £40 plus their labour contribution. The low costs were of course due in no small measure to the county council allowing the use of its water source and its quite considerable existing pipeline, and to the fact that the only paid labour was that of the carpenter and plumber.

Once the ice had been broken, as it were, the idea of group water schemes took off. In Manor Kilbride itself the first effort was quickly followed in 1959 by a more ambitious scheme involving forty-four houses and nearly nine miles of piping. At this stage the Local Government grant had been increased to £50 per house and in addition, the Wicklow county council gave supplementary grants of a similar amount.

Group schemes were welcomed by the central government and by the local authorities. Because of the voluntary labour content they were usually much cheaper than public supplies and of course they did not tie up resources needed for urgent work elsewhere. Provided that the standard of construction, pipe sizes, hardware etc., met the official specification, these schemes could be incorporated into public regional schemes at a later date. Meanwhile the householders would have enjoyed the benefits of piped water. Indeed the sense of doing something for the community's benefit was in itself a strong motivator. Desmond Roche of the Dept. of Local Government refers to the

> spirit of community development which has been their moral mainspring. Their adherence is cemented by the understandable rivalry between new voluntary and old statutory organisations. There have been signs of a tendency for the group scheme movement to turn its collective back on local authorities and all their works and to seek a kind of frontier freedom of individualism and self-reliance. The movement has been compared somewhat irreverently to the evangelical revolt from the established church.[5]

From 1956, when the first group scheme involving three neighbouring farmers in County Limerick had been installed, the ESB had been promoting this type of co-operation with a limited amount of success. Now, with the green light given by the authorities, the concept of group water supplies quickly took root in the rural community. Because of the knowledge and experience which had been acquired by REO the government requested its assistance in setting up the project. REO reacted enthusiastically. It developed a very close liaison with the various government Departments and local authorities. Its water advisory service, which had been operational on a modest scale as far back as 1947, was strengthened. It gave authoritative advice to consumers on grants, water sources, layout of systems and the type and size of pumps, piping, pressure tanks etc. best suited to cope with particular situations. In addition, using the Installation Contracts organisation in the various ESB Districts, REO frequently tendered for and obtained the contracts for the installation of the pumping and control systems for group schemes of different sizes. In this area, as with much of its installation work, the ESB saw its role as that of a trail blazer. From the earliest days of the Rural Electrification Scheme it had been involved in the installation of pumps and pumping systems and in the solution of farming water supply problems. It had developed expertise in both pumping hardware and control systems and had, over the years, kept in close touch with developments. In particular it had identified pump types and models which were especially suitable for Irish conditions. It welcomed the opportunity, whenever its tenders were successful, to introduce the most suitable equipment and to demonstrate proper installation techniques which could be observed and followed by other contractors. The introduction of each new model was accompanied by an intensive training of service electricians in its maintenance and repair and by the stocking of spare parts so as to ensure a competent after-sales service. It was hoped that if REO set high standards in the beginning other individuals and firms in the business would follow suit.

A typical group installation was that at Ballyduagh near Cashel where in 1958 six farmers dependent for all their water on a single well decided to install a pressure-storage system at the well head, piping the water from there to their dwellings and farmyards. The farms were widely separated, requiring 6,000 yards of piping, but by doing most of the heavy work themselves and by availing of the grants, the farmers kept net cost to about £110 each. This particular installation got wide publicity in the local press and set off a chain reaction of group schemes in the region.

As group water schemes were an ideal example of community development by co-operation it was little wonder that Muintir na Tíre fostered them enthusiastically. A very well publicised scheme, which set a headline for many subsequent schemes, was the Kilally and Ballinrish scheme sponsored by the Kilworth, Co. Cork, guild of Muintir. The value of the labour provided by the thirty participants was estimated at £1,200. The total cost was £3,500 and the average contribution after payment of grants was less than £17 per house, shared according to ability to pay.

The group scheme idea grew quickly and became widely accepted as a means of providing a large section of the rural community with piped water at a reasonable cost.[7] From groups of a few adjoining farmers, the schemes grew to embrace whole communities. A case in point was the parish of Killeigh near Tullamore where in 1965 the parish priest, encouraged by the success of an earlier scheme for ninety-five houses in Geashill now developed plans for supplying the other houses in his parish. REO was asked to quote for the pumps and equipment. With the closest co-operation between the parties involved, the parish priest, the parish committee, the consulting engineer, Local Government engineers, REO and of course the parishioners themselves, a group water scheme was developed to bring piped water to over four hundred houses. It was thus as large as, indeed larger than, many public water supply schemes, and more advanced in design. It had a main reservoir of 100,000 gallons capacity, six multi-stage centrifugal pumps with stainless steel impellers, guide vanes and shafts and a sophisticated electrical control system to ensure reliable and economic operation. The capacity of the well was phenomenal. Seven years later when over five hundred houses had been connected, it was calculated that over 100,000,000 gallons had been supplied from the one well without a single break.

Over the fifties, however, the ICA was not at all satisfied with the progress in providing water on tap. By 1960 when almost a quarter of a million rural homes had electricity, only about 50,000 of these had yet installed piped water supplies. On the 14 August 1960 a meeting was called by the Association at its headquarters in 23 St Stephen's Green, Dublin, with the object of launching a special campaign to develop rural water supplies. In inviting interested parties to attend, the Association had cast its net far and wide. The minutes record the attendance of Deputy Paudge Brennan, Wicklow County Council; Patrick J. Dowling, Engineer-in-Charge, Rural Electrification; Tom Finegan, NFA; Michael Owen Fogarty, Macra na Feirme; J. J. Byrne, Agricultural Institute; Noel Manahan, Unidare Ltd; P. J. Meghen, Muintir na Tíre and Limerick County Manager; C. O'Neill, *Farmers' Gazette*; Larry Sheedy, *Irish Farmers' Journal*; Joan Crady, Muriel Gahan, Nan Minahan, Phyllis O'Connell, B. Smith and D. Tomlin of the ICA. Áine Barrington was in the Chair.

The meeting set up a joint committee in the Campaign for Rural Water Supplies which over the next couple of years, in addition to widely publicising the need for water on tap, did much to co-ordinate the efforts of all parties concerned. It provided a forum for discussion on such topics as the impact of regional schemes on the rates, interim payments of grant instalments, block payments of grants in the case of Group schemes and grants for non-vested county council cottages.

The main stated thrust of the Committee's efforts, however, was 'to publicise the need for water on tap and the urgency of installing it by whatever means is found to be the most practical and economical'. It mounted conferences and exhibitions on Rural Water Supplies at An Grianán and in the Mansion House in Dublin. In all cases the fullest co-operation was given by REO, particularly in mounting the exhibitions. The Minister for Local Government indicated his support by attending the An Grianán conference and was backed up by Desmond Roche of his Department, who gave a comprehensive survey of the existing position. Fr Collins, the veteran of the Kilbride schemes, spoke of his experiences of communal 'aquafication' to an audience of over two hundred, most of them women. He finished his lecture with a rousing challenge:

> Many men in rural Ireland are not yet fully aware of the advantages of piped water supply. Very often they will get a water supply with the advantage of the stock in mind rather than the housewife, and very often if the stock don't want water, the housewife has to do without it also. This is just where you take over. Let your menfolk know you will not tolerate such treatment. Tell them you don't want to end up in your old age with a bad heart got from dragging buckets of water over long distances. Proclaim to them the gospel of rural water. Try to get them to form local groups and when they form such a group, get them to co-operate one hundred per cent in the carrying out of the work. Rural Electrification was the rural social achievement of the 1950s; let Rural Aquafication be the rural social achievement of the 1960s. Tell your menfolk what has been done in Kilbride and many areas of Ireland and say to them: 'Go and do thou in like manner'.

In its work the Committee had full and enthusiastic backing from the farming press. The *Irish Farmers Journal* in particular issued a number of supplements with contributions from experts on various aspects of rural water supply. Some of these were very extensive, as for example the twelve-page supplement issued to coincide with the An Grianán conference.

At the Turn of the Tap exhibition held in conjunction with the water conference in April 1961 a competition was held at the REO stand with a prize of a complete set of bathroom equipment, and was continued at the REO stand at the Spring Show which followed a few weeks later. Competitors, who were required to be rural residents, were asked to list in order of preference, twelve suggested benefits which water on tap would bring. Almost 8,000 entries were received and the competition was won by Mr F. Tomany from County Monaghan. The order of benefits as seen by the majority of competitors was as follows:

Saves time and labour spent in carrying buckets of water from the well
Ensures that water is available at all times at the turn of a tap
Enables a bathroom and toilet to be installed
Improves personal hygiene
Provides a safeguard in the case of fire
Makes washday easier for the woman of the house
Eases the task of preparing food, especially vegetables, and makes it more hygienic
Makes possible the installation of a hot water system
Makes automatic watering for all stock possible
With hot water, it ensures hygienic and easy cleansing of dairy utensils
Ensures healthier cattle and higher milk yields
Simplifies the problem of cleaning stock buildings.

It is worthy of note that the first eight placings referred to the standard of living in the home. The advantages out on the farm were relegated to the lower positions. This result left no doubt about the women's priorities, but it also carried some suggestion that farming was still regarded more as a way of life than as a competitive business. (This changed somewhat in the seventies. The September 1975 issue of *Prospect,* the ESB marketing magazine, quotes the example of Cloonminda, Co. Galway, where the cow population had increased by 230% and dry stock by 100% since the group water scheme became operational in 1972.)

All this activity had the desired effect of boosting demand for water on tap over the sixties. The estimated figure for electric water pumps in rural consumers' premises on 31 March 1961 was 25,000, of which 8,000 were supplied by the ESB. This gave a figure of 9% of all rural electricity consumers with water pumps. Figures given in a paper presented to the Institution of Water Engineers by Bernard J. Tighe, senior engineering inspector, Department of Local Government,[8] show that by 1966 grants for private installations were running at about 6,000 per annum. By September 1966, 173 group schemes had been completed serving 2,018 houses; 153 serving 2,579 houses, were in progress and 1,189 further schemes were pending. The costs to that date were:

Reservoir schemes: 49 schemes (906 houses), average cost £166 per house.
Automatic pressure installations: 49 schemes (355 houses), average cost £169 per house.
Extensions from mains: 75 schemes (757 houses), average cost £117 per house.

At that stage group water schemes were making water on tap available to houses at the rate of 2,000 per annum.

The regional schemes were of a much larger size. An analysis of eighteen such schemes serving 11,957 houses gave an average of 1,110 houses per scheme with a range of from 185 to 2,509. The cost per house averaged £344 with a range of from £212 to £520. Thus the group schemes cost about one half of the regional schemes per house supplied. However, the costs are not strictly comparable. The regional schemes included expensive headworks for water treatment and the ensuring of a good margin for increased demand not usually provided in group schemes and had larger pipe sizes to permit expansion. They did not, however, extend supply to the houses but provided a supply point on the public roadway to which householders could make a connection at their own expense. On the other hand, the costs quoted for a group scheme included the provision of a kitchen sink in each house, complete with cold water tap. It was intended that in the course of time most group schemes would be incorporated in regional schemes thereby ensuring a more reliable supply source and a more uniform and carefully controlled quality of water.

HOT WATER ON TAP IN THE HOUSE AND IN THE DAIRY

REO's promotion of water did not end with the provision of the cold water tap and the kitchen sink. The benefits of hot water on tap were constantly stressed. The traditional farmhouse building, however, did not lend itself easily to the roof tanks required for the conventional hot water system. A number of inexpensive electric water heaters were therefore developed to provide, in conjunction with a pressure-storage pumping system, hot water on tap at low capital and running costs. These included a self-contained cistern water heater which could supply a number of taps and which did not require a roof tank. There was also available a very basic and inexpensive non-pressure water heater with a single outlet, mainly for use at the kitchen sink.

With the increasing emphasis on high standards and the very large increase in milk production which followed Ireland's entry to the EEC, the availability of adequate hot water on tap in the dairy became of great importance. As herd sizes grew the number of farmers using milking machines increased, as did the demand for hot water to cleanse the equipment. A survey commissioned by the ESB in 1972 indicated that of the 40,000 dairy farmers having ten milch cows or more (usually the figure at which machine milking was found necessary), only about 6,000 had any form of water-heating installed in the dairy. The remaining

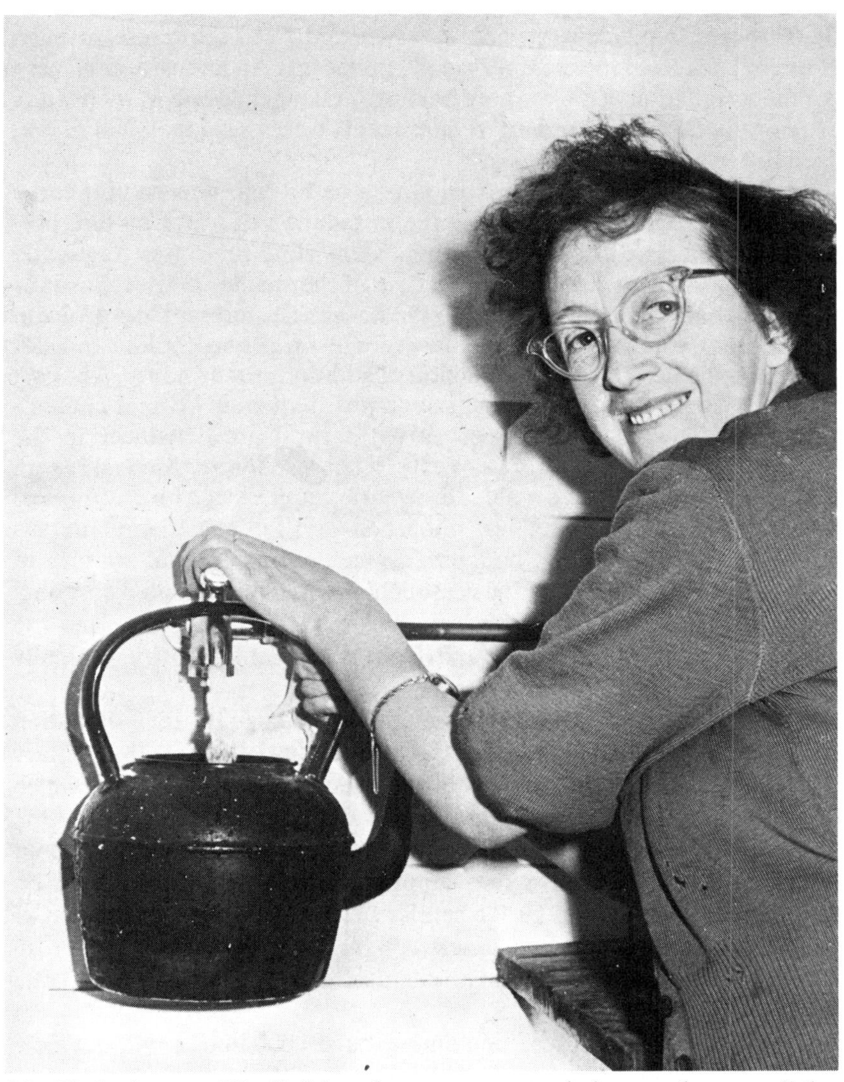

Mrs Hetherington of Errill did not have water on tap in her own home until the Errill group scheme started.

34,000 were dependent on hot water carried from the house or even brought back from the creamery in the milk cans. An intensive campaign for the installation of dairy water-heating was launched and water heaters specifically designed for dairy requirements were developed and manufactured by Irish firms.

In many houses also the provision of a bathroom, now coming to be regarded as a necessity following the installation of water on tap, presented problems. In the larger houses an existing room was frequently available for conversion, but in the case of the smaller houses it usually had to be built on as an addition to the house. This presented a problem as the do-it-yourself ability of the average rural householder did not stretch to the design and construction of additions to the house. The cost of employing a professional contractor was frequently beyond people's resources. Towards the end of 1959 the rural area engineer in the Kilmyshal area on the Carlow–Wexford border, Joseph McBride, with the help of a co-operative builder drew up a simple design for a combined bathroom and toilet which was published with full construction details in *REO News* and which undoubtedly helped many rural families to install their first bathroom at a reasonable cost. The estimated cost on a do-it-yourself effort was £226 including septic tank. At the time the grants from the government and from the local authority generally covered up to two-thirds of the cost.

Throughout the sixties and seventies the campaign for the installation of running water continued with the full backing of the rural voluntary organisations, central and local government and not least REO, which regarded its mission incomplete unless water on tap complemented the extension of electric power. In 1959 an ESB survey showed that about 106,000 or 46% of rural homes supplied with electricity had water on tap. By the beginning of 1980 a similar survey (Table 7) showed that of 425,000 rural electricity consumers 336,000 or 79% now had water on tap available through individual schemes, group schemes or the large regional schemes.

The remaining premises, including about 10,000 houses without electricity supply, thus amounted to about 100,000. About 70% of these were located in the 'west', either in Connacht itself or in Clare, west Cork, west Limerick or Kerry in the south, or in Donegal, Cavan, Monaghan or Longford (Fig. 2). As part of an overall programme for the stimulation of agricultural growth, the EEC in 1981 approved an allocation of money for the development of rural water supplies in these counties. This enabled a very generous scale of grants to be approved by the government, amounting to 90% or a maximum of £750 per house in

Gaeltacht areas and 80% or a maximum of £600 per house in non-Gaeltacht areas with special grants available for island communities.

REO Water Advisory Service, though now part of the larger Agricultural Advisory Unit of the ESB, still plays a big part in the organisation of schemes and of promotional propaganda in the provincial media. It supplies information on available grants, how to estimate water requirements, water testing and treatment, pump selection and installation, electricity supply to the pumphouse, sizes of piping etc. These features are strongly supported by advertisements from well drillers, pump suppliers and group waterworks contractors. A comprehensive booklet *Rural Water Supply – How to go about it* which gives information on almost every aspect of the problem was published by the Unit in early 1982 and supplies made available to the various authorities and the public. It can thus be seen that the thrust by the ESB for rural aquafication to accompany rural electrification, which was commenced in the late 1940s, continues as strongly as ever, as does its co-operation with the central and local government authorities to ensure that within the shortest possible period water on tap in the rural home will be as universal as electricity is at present.

Practical demonstration in pump installation by the area organiser.

TABLE 7

Rural Water Supplies in Ireland at Beginning of 1980

ESB District	*No. of rural electricity consumers*	*With water on tap*	
		No.	*%*
Athlone	37,795	25,323	67
Galway	39,974	28,781	72
Sligo	67,058	46,941	70
Tralee	33,997	23,798	70
Portlaoise	37,597	32,709	87
Dundalk	47,148	38,661	82
Limerick	48,312	37,200	77
Dublin South	5,165	4,752	92
Dublin North-west	16,331	16,331	100
Waterford	42,927	37,776	88
Cork	48,999	43,609	89
Total	425,303	335,881	79

Source of Household Water Supply

	%
Piped Water	
Public water mains	33.9
Group water schemes*	15.4
Own electric pump	28.6
Miscellaneous	1.5
	79.4%
Other Sources	
Roadside tap or pump	4.7
Well	15.9
	20.6%

(ESB Consumer Survey, 16 September 1980)
*From information supplied by Department of the Environment at end of 1981, about 50% of Group Water Scheme consumers are supplied from extensions to public mains and 50% from own sources, the latter systems almost always involving electric pumps.
Note: The 100,000th consumer in a group water scheme was connected in October 1982.

CHAPTER EIGHTEEN

Electricity on the Farm

WHEN IT CAME TO DEVELOPING the use of electricity in agriculture, the small size of the Irish farm, the limited cash resources of the average farmer and the unsatisfactory markets for his produce were inhibiting factors. A comparison with Britain made in 1956 (Table 8) showed that, counting only farms of over five acres, 46% of British farms were over fifty acres, as against only 28% in Ireland.

The cheap food policy of Great Britain, until the late 1960s the only export market of any consequence open to the Irish farmer, held prices down. The British farmer was compensated by various grants and subsidies which were not of course available to Irish farmers. In contrast also to the British situation of considerable paid labour was the fact that in Ireland most of the farms were family units with very little paid help. Mechanisation in the farmyard would in such cases appear to save the farmer little money and such investment as could be made out of meagre resources was generally in land improvement or possibly in the purchase of stock or a tractor. There was of course the odd farmer who was quick to appreciate the advantages of farmyard mechanisation, and who was in a financial position to avail of these; but in general up to the mid-1960s the development of electricity in the Irish farmyard was much slower than in the farmhouse.

From the mid-sixties on, with the prospect of the opening up of European and other foreign markets, a change was discernible in the style of Irish farming. The mixed farm, with its 'little bit of everything' began to give way to the specialised farm with one or two strong enterprises, dairying, sheep, tillage, poultry, pigs etc., with a far greater output than hitherto. This style and intensity of operation, displaying as it did a more scientific and commercial approach than formerly, placed far more emphasis on labour saving and productivity and consequently on electrification. There were thus two distinct phases of farming style and activity encountered by REO with the watershed occurring somewhere around the middle 1960s. These two periods presented different challenges and problems.

TABLE 8

Number and percentage of farms of over 5 acres in each size group in Great Britain and Ireland 1956.

Size	*Great Britain*		*Ireland*	
(Acres)	*No.*	*% of Total*	*No.*	*% of Total*
5-15	91,000	25.8	60,776	21.0
15-50	99,000	28.0	147,986	51.0
50-100	69,900	20.0	51,755	17.9
100-200	50,160	14.6	21,928	7.6
Over 200	41,420	11.6	7,249	2.5

Source: Paper, 'Rural Electrification in Ireland' given to 1956 Annual Conference of British Electrical Development Association by P J Dowling, BE, BSc, ARCSc I.

THE EARLY YEARS — 1946–65

Irish farming at the end of World War II was not in a healthy state. While the agricultural advisory services and rural vocational schools made trojan efforts to raise farming standards they had to contend with a serious lack of resources and a considerable apathy among the farming population. In every county, however, a small number were prepared to experiment and it was chiefly among these that REO concentrated its promotion and development. One of the most promising areas was in grain processing, conveying, drying, ventilated storage, grinding and rolling. It was considered that a 'grow your own, process your own, feed your own' policy with regard to feeding stuffs would reduce considerably the feeding costs of livestock and permit profitable expansion. With the help of electricity, the handling, drying and processing of home-grown grain could be carried out at a reasonable cost and without cutting in unduly on the farmers' time.

Grain Milling and Grinding □ Among the first farmyard appliances offered in 1947 was a 3hp plate mill of Canadian manufacture with a built-in motor which had a take-off pulley for driving other appliances, such as rootpulper, small circular saw, grindstone or milking machine. It cost £45 when introduced and had an output of up to 3 cwt per hour, depending on the fineness or coarseness of the grind, for three units of electricity. It was at the height of its popularity in the late 1940s and early 1950s when increased costs and dollar difficulties caused REO to look for a cheaper substitute. Nevertheless, the Canadian Woods grinder continued to give good service. In 1980 a farmer in East Donegal reported that his Woods grinder was still working happily. After thirty years he had found that replacement plates were still available from the ESB.

The substitute for the Woods grinder was the German 'Bauerngold' (literally 'farmers gold' and rapidly anglicised to 'Barngold') which, instead of steel plates, had grindstones of a quartz-concrete mixture. Powered as it was by a 1hp motor, it had only one-third the output of the Woods and so took three times as long to produce the same amount. As the minimum amount of attention was necessary, this caused no problem: the farmer could leave the machine running and attend to other tasks. The amount of electricity used per hundredweight of meal ground was about the same as for the larger machine, about one unit.

This illustrates a message which REO was constantly preaching to farmers. For the average or small farmer, there was no need to invest in large machines where electricity was concerned. Taking the grinder as an example, if the farmer's requirements were, say, 6 cwt per day of ground meal, the electricity cost was much the same whether it was produced by a 3hp machine in two hours or by a 1hp machine in six hours. Provided hoppers and bins were of adequate size, the smaller machine could operate without attention and, if provided with a simple pressure switch, could be made to switch itself off at the end of the operation. Grinding by tractor, on the other hand, required constant attention and so a large machine was desirable to reduce the time taken to a minimum. The same applied to pumping. A ¼hp motor could pump 250 gallons per hour from the average well continuously without any supervision. This added up to 1,000 gallons for four hours for one unit or 6,000 gallons in twenty-four hours for four units of electricity at a cost of less than 6*d.* (2½p). In the case of the Barngold mill, the message was well taken, as, despite its small capacity, it rapidly became a best seller. Thousands were installed throughout the length and breadth of the country.

For some purposes, such as poultry-feeding, a hammer mill was considered more suitable, as it broke up the husk of the grain more effectively than did the stone or plate mill. The grain, having been broken up by the hammers, had to pass through a screen or sieve which could be as fine or as coarse as required. Thus, a far more precise control of the product could be achieved. Imported hammer mills were rather expensive, but in 1954 an Irish hammer mill was developed by Patrick J. Tobin who had a small engineering works in a truly rural setting at

Edermine near Enniscorthy, Co. Wexford. This was a good example of the feed-back effect of rural electrification and the opportunities it afforded to native enterprise. The mill was powered by a 3hp motor and had an output of from 2 cwt to over 12 cwt per hour, depending on the fineness of the screen and the type of grain being ground. The Tobin mill proved a most popular machine among Irish farmers. At a later stage, a similar, but smaller, hammer mill of 1½hp was produced by ACEC of Waterford.

Grain Drying □ Grain drying by electricity on the farm was proving quite successful in Great Britain, but the average Irish farm needed a smaller dryer. For the medium to small grain growers, particularly those who wished to store and feed their own grain a series of small tray-type grain dryers was developed by REO. These were on the do-it-yourself basis, REO providing the fan-heater units and complete plans for the building of the dryer. They were of three sizes, with heater units of 5kW, 10kW and 15kW, which were capable of a moisture extraction of 6% from ½ ton, 1 ton and 1½ tons of grain, respectively, in 10 hours. While some hundreds were installed, their low throughput and labour intensiveness inhibited their popularity, particularly in later years, when the acreage of barley grown had increased greatly.

Grain Conditioning □ For short-term storage, pending delivery to the grain merchants, a system of grain conditioning was advocated which involved the quick cooling of the grain (i.e. within 10 days), to 10°C by ventilating it with cold night air. This system ideally suited the producer in the 30-ton range, since a single 1½hp fan would achieve this target. Investigations carried out by An Foras Talúntas indicated that green wheat with an initial moisture content of 21% could be stored without deterioration for up to six weeks by this method.

Grain Handling □ For handling grain, a very simple screw or 'auger' grain elevator powered by a fractional hp motor was very popular, eliminating as it did much of the back-breaking shovelling associated with grain movement.

Equipment Prices □ Over the thirty years of the scheme, there were of course many price changes, usually upwards. In order to give some idea, however, of prices, those ruling for the most popular items in August 1960 (about half-way through the scheme) are given, including motor and starter where these were separate from the appliance:

Above: Grain milling and grinding — the Woods 3HP grinder with auxiliary take-off pulley. *Below*: Safe from crushing by 'mother' in their electrically heated creep.

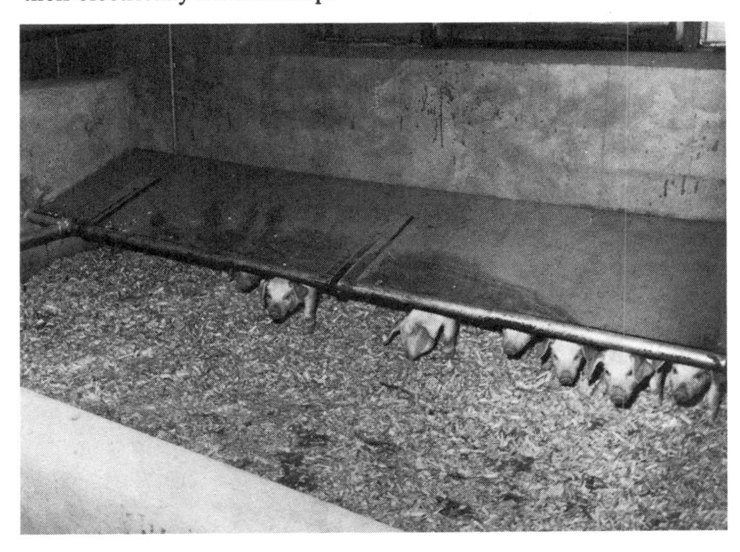

Barngold mill	£49.00.0	
Barngold oat-roller	£60.00.0	
ACEC hammer mill	£59.10.0	
'Tobin' hammer mill	£81.00.0	
Bentall automatic hammer mill	£129.07.0	
Mayrath auger elevator	£28.10.0	depending of length
	£30.00.0	
	£35.00.0	

Grain drying fan heaters	5kW	£35.10.0
	10kW	£46.12.0
	15kW	£65.15.0

For the larger farms more sophisticated equipment was available, limited only by the depth of the farmers purse and the economics of its adoption. The larger the throughput and the wage bill, the stronger the case for mechanisation and automation. In the early years of the scheme, however, these farmers were very much in the minority. Once the EEC markets opened up in the 1970s, a considerable number of farmers, or co-operatives of farmers, engaged in specialist production, which involved much larger units than those described.

The Farm Workshop □ Deputy Jim Hughes, one of the TDs who was an enthusiastic advocate of rural electrification at its very inception confided to P. J. Dowling, when the latter was Engineer-in-Charge, that of all the possible benefits of the scheme, the one which most attracted him initially as a farmer was the prospect of an electric grindstone to keep his tools sharpened.

Rural electrification coincided with increasing mechanisation on the farm. New and second-hand tractors flooded in from Britain after the war and electricity helped farmers to carry out much of their own maintenance and repair. One of the most popular appliances was the electric drill with its many attachments and one of the larger drill manufacturing companies conducted a week-long factory seminar for REO salesmen. Another versaatile appliance was the single-phase electric welder, which sold for about £50–£100, depending on size. With a drill, a welder and a few simple hand tools, the farmer could not alone carry out extensive machinery repairs, but also fabricate many items of his own design for use on the farm or around the farmyard.

The rural blacksmith also received a new lease of life just when he needed it. Horses were disappearing as mechanisation increased, but with an electric fan for his fire, an electric welder and a small range of powertools, the blacksmith was now into a new and fast-expanding business in the repair of farm machinery.

It was hoped that the ¼hp or ⅓hp motor (selling for about £10 and using only one unit in 3 to 4 hours), combined with a self-assembly speed reduction unit (costing about £8), would prove a popular and inexpensive introduction for the smaller farmers to the cheap motive power now available for mechanisation of many farmhouse and farmyard tasks. The slogan was 'Whenever there is a handle to be turned, electricity can do it better and more cheaply'. In particular, it was hoped that it would be applied on a large scale to such regular chores as churning in the farmhouse and root-pulping in the farmyard, both well within the capabilities of the small motor.

However, while some consumers were receptive, hopes for general acceptance of even such a modest level of mechanisation were not realised. The only really wide-scale use of the fractional hp motor in the early years was for water-pumping, generally as an integral part of a pressure-storage system.

REGENERATION – THE SIXTIES

By the end of 1958 there were signs that a fresh breeze was preparing to blow away the economic doldrums of the fifties, demonstrated by contrasting extracts from the *Irish Review and Annual Supplement* published by *The Irish Times* for the years 1957 and 1958. In the article on the economy for 1957 we read that 'Production continues to languish, emigration to flourish. . . . Emigration is attributed . . . to economic backwardness and our economic backwardness is due to lack of productivity . . . the cattle trade is faced with imminent disaster, as more and more areas in Great Britain are closed against beasts that are not free from tuberculosis'. Later we read of a decline in the export of eggs and poultry.

In contrast, in the issue for 1958, the Economic Survey opens thus:

> A bewildering succession of new developments marked the year 1958 in the Irish economy. New events and new ideas crowded the scene and a new spirit, very different from the dejection of 1956 and most of 1957 motivated the actors . . . agriculture, industry, external trade, finance and the Government sector were all affected in varying degrees by the quickening tempo of economic activity and the recovery of confidence in future prospects.

Much of the stimulus for this change was provided by the publication by the government in mid-November 1958 of a White Paper, 'A Programme for Economic Expansion', based on a study by T. K. Whitaker, Secretary Department of Finance, entitled 'Economic Development'. The programme emphasised that increased agricultural production would for the most part be for sale abroad and must, therefore, be competititve. 'State assistance should concentrate less on price supports and guarantee than on measures designed to bring about increased agricultural productivity, that is lower production costs per unit of output. It is the only way that will get increased agricultural production and achieve incomes on a sound and permanent basis.' The programme went on to analyse the problems of the different sectors of agriculture and outline a strategy for dealing with these.

While the major advances in the rural economy were still some way into the future, it could be said the 1958 marked the beginning of a change in attitude which resulted in the upsurge in agricultural productivity and incomes, which was such a strong feature of the late 1960s and particularly the 1970s following Ireland's accession to the EEC. This was reflected in the growing use of electricity as an aid to production. On the other hand there is little doubt that without electricity Irish farmers would have found it impossible fully to exploit the opportunities provided by the opening up of the Common Market.

Much of this new dynamic in Irish agriculture was provided by the foundation, also in 1958, of An Foras Talúntais (The Agricultural Institute). The Institute was set up following an agreement between the governments of Ireland and the United States and financed out of Counterpart Special Account money provided by the United States under the Marshall Aid Programme 'to review, facilitate, encourage, assist, co-ordinate, promote and undertake agricultural research, including horticulture, forestry and bee-keeping'.

Prior to the setting up of the Institute, facilities and funding for agricultural research were very limited. Research into and development of the application of electricity in agriculture were almost non-existent. It was obvious to REO from experience in other countries that electricity had much to offer agriculture, but it was equally obvious that electrical applications which were successful elsewhere were not going to be simply and easily transferred. In the 1950s a limited amount of work was done by REO on certain applications of electricity to agriculture and horticulture with the enthusiastic help of the Agricultural and Horticultural Departments of University College, Dublin. These mostly involved the testing out under Irish conditions of techniques developed as a result of basic research elsewhere. The Agricultural Department carried out a programme of experiments in environmental control for stock-rearing, crop drying and barn hay drying by electricity. A parallel horticultural programme also involved environmental control. This included potato sprouting, soil warming, plant irradiation, supplementary lighting, growing rooms for tomato plants, automatic watering and mist propagation. Resources for basic research or for the widening of the scope of the experiments were, however, very limited.

The setting up of The Agricultural Institute, with its vastly greater resources, permitted a much wider research programme into the applications of electricity to agriculture and horticulture. REO worked in close co-operation with the Institute in the appropriate programmes. The research work of the Institute was not of course confined merely to agricultural technology. It also embraced agricultural economics, farm management, marketing, farm buildings and rural sociology. The results of its wide spectrum of research were transmitted to the farming sector through the County Committees of Agriculture and the local agricultural advisers. Before long a new and more enlightened approach to farming and its problems was discernible. This was helped by the educational work of the voluntary organisations and by the increasing availability of second-level education. One of the great changes in the case of many small farmers was the transition from the old system of mixed farming into more profitable specialisation patterns which suited their particular circumstances — dairying, pig breeding, poultry production, etc. As the pattern changed and with increasing emphasis on efficiency and productivity, there was an increasing usage of electric power. As discussed elsewhere, one of the big problems facing the ESB in the late 1960s and 1970s was reinforcement of the rural electricity networks to meet this rapidly increasing demand.

Milk Production □ One of the greatest contributions of rural electrification to agricultural production has been in the dairying industry. From 1970 to 1979 milk production increased by 56%, from 640 million to 1,000 million gallons. About 85% of this went into manufacturing, mainly butter and cheese of high quality for which markets had opened up in the EEC. Butter exports increased from £10.6 million in 1970 to £198 million in 1979 and cheese exports from £6.2 million to £85 million. In the same period the agricultural work force dropped by 22% from 276,000 (26% of total at work) to 214,000 (18% of total).

This increase in milk production was due not only to higher yields per cow (the number of cows had increased in the same period by only 23% — from 1,713,000 to 2,208,000), but also to the higher productivity in the farmyard and in the dairy by a much smaller work force. Well-designed herd handling and milking methods using milking machines permitted one-person milking of even very large herds. Electrically refrigerated milk coolers and bulk milk tanks permitted quick cooling, longer storage periods, higher quality milk and more cost-effective collection methods. Electric pumps ensured adequate water supply for drinking, pre-cooling and washing, while specially designed electric dairy water heaters with automatic time-switch control provided adequate hot water for personal cleanliness, udder washing and cleaning of milking equipment, thus enabling the producer to meet the exacting hygienic standards required by modern markets. Specially designed electric pumps enabled the high-pressure hosing down of byres, milking parlours and yards, also helping to keep up high standards of hygiene required on the modern dairy farm.

The number of milking machines grew from about 1,000 in 1946 and 10,000 in 1960 to an estimated 60,000 in 1979. The number of milk coolers, including refrigerated bulk milk tanks, grew from about 30,000 in 1970 to 46,000 in 1979, the increase being mainly in the refrigerated bulk tanks.

Pig and Poultry Production □ Modern living styles and marketing requirements have ensured the demise of the traditional picture of the pig as 'the gentleman who pays the rent' in practically every Irish farmstead. The tendency is more and more towards a concentration of pig production amongst a smaller number of producers. In 1970 and 1980 the number of pigs produced was about the same, at about 1.1 million, but the number of producers had dropped from 24,000 to about 10,000. Some of these are very large producers — an article in *The Irish Times* in March 1982 quotes twelve large units in County Cavan with some 10,000 pigs or more each. With such large concentrations of animals, precise control of the environment is vital for optimum growth and prevention of disease. Electric fans for ventilation, electric floor heating and infra-red lamps in creeps and farrowing pens, electric pumps for slurry disposal and for washing down, all contribute to efficient production and hygienic conditions.

Poultry production has followed the same lines as that of pigs, larger concentrations of birds requiring similar control of environment. With such large numbers, milling and mixing of feed can be most economically carried out on the premises and again the availability of electricity permits efficient production with the minimim of attention. For the very large units, mechanised feeding equipment has been developed, which

Above: Effectiveness of electric power for cleaning at an early morning demonstration. *Below*: Hot water for washing the udder and for stimulating the 'let down' of milk preparatory to the use of the electric milking machine.

greatly increases the number of animals or birds that can be looked after by one person.

Waste Management and Cleaning in the Farmyard □ The intensified production methods on the farm and in the farmyard have brought with them their own sets of problems, particularly in the area of waste disposal. In the traditional small farm, with straw bedding for comparatively few animals, the disposal of waste first to the farmyard dung heap and then on to the land was a manageable task for the farmer and was a system which had provided an ecological balance over thousands of years.

New and more intensive methods of rearing, feeding and housing stock necessitated by larger animal numbers and a smaller work-force, increased areas under concrete and increased quantities of silage have given rise to new problems of effluent disposal. The effluent waste in dairy herds from milking sheds, parlours and adjacent areas consists of solids, soiled water, detergents, seepage from dungsteads and silage pits and all waste water that has been in contact with animal manures in calving boxes, wintering sheds etc.

Intensive strawless housing means that in most cases to-day, animal manure is slurry, a semi-liquid material. The amounts involved are large. A herd of a hundred cows or the same number of 500-kilo beef cattle will produce 900 gallons of slurry per day. Two problems immediately arise — the removal of the slurry from the farmyard and buildings and its disposal in such a way as not to cause pollution elsewhere. The fertiliser value of this slurry can be quite high; for example at 1980 prices its use on a two-cut silage could save up to £25 per acre in replacement or artificial fertilisers. On the other hand, its excessive use, particularly in drumlin country, where it slides off with rain into the rivers and lakes, can be most harmful to fish life.

Electrically operated units are available to deal with all kinds of slurry pumping whether it is from underground to overhead tanks; from storage tank to tanker/spreader or direct to the land by sprinkler systems. Some pumps are equipped with cutting mechanisms for dealing with foreign matter (straw, twigs etc.) in the main storage tank. The power required varies but many units are installed with motors as low as 3 to 5hp.

Grain Storage □ With the more intensive stock rearing the amount of barley grown for home feeding increased from 335,000 acres in 1968 to 645,000 acres in 1978.[1] While the small tray-type electric dryers developed during the 1950s and early 1960s had not been very successful, chiefly because of the amount of handling involved, a more promising

In the farm workshop the electric grindstone is vital for keeping tools sharp.

method of storing comparatively large quantities was developed by An Foras Talúntais in co-operation with the ESB. This system was based on research carried out by the British Electricity Council and, in contrast to previous systems using both motive power and heat, used motive power only to blow cold air through a holding bin. It was much more economical than those involving heat. It enabled barley to be held safely for home use for up to six months. Because of its cheapness and simplicity it was widely adopted by even the smaller farmers. By 1976 about 10,000 installations of this type of drying/storage unit were in operation.

ELECTRICITY IN HORTICULTURE

Before the advent of An Foras Talúntais a notable amount of research and experimentation was carried out by Professor Edward J. Clarke and Dr Joseph V. Morgan of the horticultural department of University College Dublin in conjunction with REO. The earliest experiments were in the area of electrical soil warming for hotbeds and bench warming in the glasshouses for the earlier production of plants from seed and propagation from cuttings. In the case of the latter the addition of artificial misting using high pressure pumps and electronic control gave very positive results and the system was quickly adopted by commercial growers.

Experiments on supplementary lighting for the earlier production of tomato plants were also very successful, and were followed by the development of growing houses for commercial production of plants. These gave, in a very small insulated house, a completely controlled environment with lighting, heating and watering controlled so as to provide ideal growing conditions with the minimim of energy input. This technique was also quickly adopted by commercial growers to produce strong early plants and thereby early tomatoes, which earned the highest prices. With the establishment of the Kinsealy Research Centre of An Foras Talúntais, research in these areas continued on a more intensive scale, again with close co-operation from REO. Automatic watering and ventilation were also subjects of research and experiment which helped the Irish tomato industry to expand rapidly in the 1960s and early 1970s. The oil crisis of 1973 and the equally serious but less dramatic increase in oil prices in the late seventies have, however, had a disastrous effect on this energy intensive industry. In 1973 the total cost of the energy inputs per acre was under £4,000. In 1978/79 this cost had risen to £14,000 and in the 1979/80 season it approached £20,000.[2]

At present the industry is vigorously seeking methods of reducing its high energy costs. Two possible fields of investigation in which electricity might help are in the use of heat pumps and of extract heat from electricity generating stations.

Heat Pumps □ By upgrading heat extracted from the outside environment, a river or a lake, a heat pump can supply as heat up to four or five times as much energy as is required to drive it. High capital costs, however, at present inhibit the use of heat pumps for overall glasshouse heating. Nevertheless, experiments in Kinsealy indicate that a combination of root zone warming (possibly by heat pump) and lower air temperatures in the glasshouse might be successful in reducing costs without undue loss of crop.

Extract Heat from Electricity Generating Stations □ While not strictly in the ambit of rural electrification, a short mention of an experiment in Lanesboro, Co. Longford, which uses heat extracted from the turf-fired generation station is relevant. By means of a modification of the boiler plant in the station, steam is extracted at a pressure and temperature higher than that at which it is normally condensed. This steam then heats water in a heat exchanger to a temperature of about 90°C similar to the temperature of water in a normal glasshouse heating system. In Lanesboro, the hot water is supplied to a pair of glasshouses of one acre each, growing commercial tomato crops. The costs (1981) of the heat supplied came to about 60% of the cost of providing an equivalent supply of heat from an oil fired boiler. With future increases in oil costs the percentage saving to the growers using the extract heat is expected to increase. The total heat supply being provided is 3.7 MW of which the loss of electricity production is 0.53 MW. Allowing for the fuel needed to replace this electricity at another station, the coefficient of performance (ratio of energy supplied to energy consumed) for the scheme is approximately 2.4 to 1.

The progress of this experiment is being monitored very closely by its sponsors which include the government, the Bank of Ireland and the ESB, as well as by the horticultural industry as a whole. Experience will show whether a solution to the present high energy costs in glasshouse cropping may lie in this direction.

Potato Sprouting by Electricity □ The practice of sprouting seed potatoes by natural light goes back many decades if not centuries. Sprouting by artificial illumination was first tried in Holland about 1946. It rapidly spread to the neighbouring countries of Europe. The idea was taken up in Britain in 1954 and in Ireland in 1961.

Seed potatoes which are exposed to adequate light and protected from frost during the winter develop short sturdy sprouts which ensure a fast growth when planted resulting in heavier yields of both early and main crop potatoes. The planting can be more flexible to take advantage of better weather conditions and the more advanced crops suffer less from potato blight. Properly sprouted seed can be used in mechanised planters without risk of damage, crop growth is more even and there are few missed spaces — as often happened when unsprouted seeds failed to germinate.

Using fluorescent electric lamps to provide the lighting allowed the use of existing well-insulated frost-free houses, rooms or cellars instead of the expensive glass-walled houses required for sprouting by natural light which normally required some heating in frosty weather. Increases in yields of 50% to 75% were reported. While the technique is now well established it was again co-operation between REO and UCD which enabled it to be successfully launched. REO acquired, stocked and sold the requisite fittings and gave instructions on site on their proper use.

This chapter has endeavoured to trace, mainly by examples, the development of electricity as an aid to Irish agriculture and horticulture. The lack of adequate research, experimental and advisory services and of adequate markets was in the early years of rural electrification an inhibiting factor. In recent years the farming picture has changed dramatically. The more scientific and commercial approach now existing within the industry and the better marketing opportunities have been reflected in the better use of electricity as an aid to agricultural productivity thus realising the hopes of the sponsors of the Shannon Scheme in 1925 and the Rural Electrification Scheme in 1946.

CHAPTER NINETEEN

Rural Industries

IT WAS HOPED that in addition to providing the means to a better standard of living and increasing agricultural productivity, rural electrification would enable industry to come to the smaller centres of population. This applied particularly to the smaller industries and indeed to the individual craftworker. The Report of the Commission on Emigration echoed this hope: 'It [rural electrification] could ease the way for the return of the craftsman to rural areas, providing employment for him both in agriculture and in cottage industries.'[1]

Possibly the first small industry in rural Ireland to result from rural electrification was in the village of Bansha. In March 1948, work started on the electrification of Bansha Area, the first in County Tipperary. Simultaneously, work also commenced in the village on the conversion of an old mill to a jam factory. The concept was that all the fruit grown by the farmers of the parish would have an outlet in the factory which would thus, in addition to providing employment in the village, provide an incentive to grow more and better fruit. Under the leadership of Fr Hayes the local guild of Muintir na Tíre formed a limited company with most of the parishoners as shareholders. Unfortunately for a number of reasons the undertaking did not survive. The next small industry recorded as having started up (also in 1948) was a weaving mill in Kilsallaghan, Co. Dublin, followed in early 1949 by a flax scutching mill in Tydavnet, Co. Monaghan.

Slowly, as rural electrification spread through the country, small industries emerged. Very often these were an extension of existing businesses. For instance the local garage or the local blacksmith installed a welder, lathe, grinder, drills and perhaps a power hacksaw and developed as a general engineering works. The small engineering business of Patrick Tobin, Edermine, Enniscorthy, was a case in point. When electricity became available, Mr. Tobin installed an electric welder and became a sizeable manufacturer of farm gates, cattle crushes and similar agricultural equipment. In 1954, with the help and encouragement of REO, he developed a hammer-mill which could be sold at an attractive price. Over the next decade the number of 'Tobin' mills sold to Irish farmers reached four figures. Examples like this began to emerge on an increasing scale as rural electrification progressed, but the big boost to rural industries came with the availability of grants and other aid from two state organisations — Gaeltarra Éireann and the Industrial Development Authority.

GAELTARRA ÉIREANN

Gaeltarra Éireann was established in 1928 to improve the organisation, conduct and development of the rural industries in the Gaeltacht regions, i.e. regions where the Irish language is still spoken as the vernacular. In this it was the successor of the Congested Districts Board established in 1891, which among other things was charged with the task of generating economic activity and alleviating the extreme poverty then prevalent in the west of Ireland where most Gaeltacht regions are situated.

The overall jurisdiction of the organisation covered parts of counties Donegal, Mayo, Galway, Meath, Kerry, Cork and Waterford, with a population (1979) of about 77,000. The Gaeltacht areas consist generally of scattered, isolated and physically remote rural areas, with no industrial tradition. Poor drainage exacerbates the problems of the lime-deficient, windswept soil. No local wealth-base exists to fuel enterprise in either industry or trade. In terms of overall natural resources these were the poorest areas in the country traditionally exhibiting the highest unemployment and emigration rates in the state.

From its foundation, Gaeltarra Éireann was empowered to establish rural industries. The advent of rural electrification opened up new possibilities in this regard, even in the most remote localities. Because of their remoteness, most of the Gaeltacht areas were well down in the queue for electricity supply. In the early fifties, however, the names of these areas appear with increasing frequency in REO reports. In 1965 the powers of Gaeltarra were extended to allow the formation of joint ventures with private industry and to grant-aid new projects. In 1971 the limits on these were broadened to allow for the exercise of full-scale industrial development programmes. These new powers, coinciding with the now general availability of electricity throughout the Gaeltacht areas, heralded an upsurge in the development of industries in these regions.

One of the most impressive achievements of Gaeltarra Éireann was the establishment of a modern industrial estate in the Gweedore district on the west coast of Donegal. This thickly populated locality, devoid of

Above: Electricity provided the impetus for agriculture-based industries as well as for new manufactures. *Below*: Giving advice on farm equipment from the mobile showrooms.

natural resources apart from the sea and the bogs, had a long tradition of emigration and migration. The availability of electricity in 1954 opened up new prospects, which were eagerly seized by Gaeltarra and by the people. There is now a large industrial complex, which gives good employment to about six hundred local people and which includes the manufacture of sophisticated communications equipment as well as radiators, packaging materials, food processing and worsted yarn. In contrast to the small pole-mounted 5kVA single-phase rural transformers of 1954, a large substation of 7,400kVA supplied from the 38,000 volt grid is now required to meet the electricity demand. Gweedore is now a thriving community with hundreds of new houses with modern amenities. Many of these have been built by former emigrants returning to work and rear their children in the place of their birth. The population, which was declining steadily over the fifties and sixties, rose by 14% from 1971 to 1979.

Not all the industrial achievements of Gaeltarra Éireann, and its successor Údarás na Gaeltachta, have been on such a large scale. The size of industry initiated directly by the Údarás or encouraged by the provisions of equity or grants, varies from the enterprise employing several hundred to that employing only one or two, perhaps a husband and wife. The cumulative effect, however, has been striking. By 1979 the number of subsidiary, associate and larger grant-aided companies totalled eighty-six with a total annual sales of £36 million. The total factory area amounted to 170,000 square metres and the number of persons employed in that year in industries sponsored by the Údarás was 4,500, a very significant contribution in the context of a total population of 77,000.

The sophistication of some of these modern industries is impressive. In addition to the indigenous type of industry such as fish processing, vegetable freezing, boat building, spinning and weaving, there are computer services, manufacture of optical lenses, hardened glass and glass fibre, quartz crystal, medical products, sports garments, fitted kitchens, air conditioners, industrial processings and engineering, photographic and printing services.

The availability of local employment in industries such as these and the amenities made possible by electricity have helped in a major way in the rebirth of hitherto depressed and isolated areas. In no sphere is this transformation more manifest than in housing. The picturesque thatched cottages have well-nigh disappeared and have been replaced by bungalows in the modern style. They are frequently criticised (and some with justification) as being too suburban in appearance and out of sympathy with their environment. Be that as it may, they are comfortable, easily run and equipped with essential amenities including bathrooms, modern sanitation and in many cases central heating. Judicious landscaping and planting of trees and shrubs will help blend them with their surroundings.

Industrial Development Authority □ The small industries programme was launched in 1967 by the Industrial Development Authority as a pilot scheme to develop a programme of support and advisory services for small manufacturing firms in Ireland. It was rapidly extended to cover the whole country outside the Gaeltacht areas (already catered for by Gaeltarra Éireann). Other IDA programmes promoted the setting up of medium and large industries and, while some of these were situated in rural locations, it could not be said that many owed their existence to rural electrification. The objective of the small industries programme, however, was to aid existing small firms and to generate new small industries and provide employment opportunities especially in small towns and villages which would be unlikely to attract other types of development. Rural electrification had opened up practically every rural village and hamlet in the country to aid under this programme. Between 1 April 1967 and 31 December 1977 the programme provided grants to thirteen hundred small firms, spread throughout the length and breadth of the State amounting to £11.7 million out of a total commitment of £23.3 million. The total employment on the latter date was 16,800. This impressive growth was stimulated not only by the grants available but by the readily accessible advisory service provided by the IDA and, particularly in the case of the western counties, by the initiative of the county development officers and the county development teams.[2]

Again, as in the case of the Gaeltarra industries, the variety of enterprises is prodigious, demonstrating the versatility of the rural entrepreneurs. They cover all aspects of engineering, from precision work to machinery repair, textiles, clothing, knitwear, boatbuilding, furniture, plastic products, coachbuilding, craft products, electrical, agricultural and mining equipment, fish processing, pottery, plastic products, steel saws, moulding equipment, marine engineering, glass crystal, hardwood veneers, liquid crystal displays and fibreglass products. A couple of examples are given in some detail.

1. *Moffett Engineering Ltd., Clontibret, Co. Monaghan* □ This industry is located in a completely rural setting about seven miles north-east of Castleblaney and four miles from the border with Northern Ireland. Its managing director is Carol Moffett whose grandfather in his time operated the local smithy. Her father, Cecil Moffett, who inherited the

business, recognised the opportunities in agricultural engineering and expanded into the manufacture of ploughs, sprayers and similar farming items. In 1954 rural electrification came to Clontibret and enabled him to widen his range of product. In *REO News* October 1958 mention is made of a 40hp grain mill in County Monaghan manufactured by Moffett of Clontibret. He required a good steel-cutting saw but as those on the market were very expensive he designed and manufactured his own. This was so successful that he commenced the commercial manufacture of steel-cutting friction saws, which were marketed under the trade name 'Monacut'. A tractor-mounted hedgecutting saw was added to the range of products and also marketed successfully.

In 1972 when Carol, aged 19, was a student of languages at Trinity College, Dublin, her father died and she, being the eldest, had no option but to return home and take over the business which employed three people. She recognised that if it was to prosper under modern conditions, a more sophisticated engineering input was needed. Up to this, the main operations had been shearing, bending and welding of steel in various forms. Now she added hydraulic and electric drives and controls and embarked on a much wider product range. With the help of the county development team and a grant from the IDA a new factory premises of 24,000 square feet was built at some little distance from the original forge and workshop. Thirty-three local people are now employed in the manufacture of a wide range of engineering equipment including the Moffett hydraulic 'Steelmaster' punching and shearing machine and the most ambitious development, the Moffett 'Multicast' production system, which is a very sophisticated automatic plant selling at £70,000, for the production of a wide range of precast concrete products such as post and panel fencing, reinforced panels, insulation panels, roadside fencing, security fencing, slatted flooring for cattle houses and vineyard poles. For her markets Ms Moffett has looked to the continents of Europe, Africa, South America and Australia, travelling widely in search of customers rather than waiting for the customers to seek her out, and availing to the full of the advice and assistance provided by Córas Tráctála Teo. Her pace of working normally requires a fourteen hour day.

In 1981 she sold two large Multicast systems to a Swedish company against strong competition. The purchaser had seen her first Irish installation in Daingean, County Offaly. Before the deal could be completed the machine had to meet the very strict requirements of the Swedish Bureau of Standards. Moffett Engineering Ltd. had to guarantee that the machine, which had to be delivered (in nine large trailer loads) erected and handed over in working order would turn out a product continuously every six minutes. A factor which clinched this particular order was that the Moffett machine could produce units of up to six metres in length, whereas the largest generally available on the market at the time could only handle four metre units. Later came the Moffett 'Maxigrip', a clamp for mounting on fork-lifts and delivery trucks for the loading and unloading of blocks, bricks, paving-slabs etc.; the 'Multicore' system for the production of concrete pipes and manhole rings; and the 'Mixveyor', a conveyor which extends the off-loading ability of a ready-mix concrete truck.[3]

2. National By-Products Ltd., Castleblake, Rosgreen, Cashel □ In complete contrast to the first example is an industry founded, again in a completely rural setting, by the Ronan brothers near Rosgreen, Cashel, a village probably better known for its proximity to the famous Vincent O'Brien stables. In 1958, following the coming of rural electrification, the brothers Louis and John Ronan took on the development of a large-scale piggery. Most of the feed came from swill and offal from neighbouring towns and meat-processing plants. As the brothers acquired a greater knowledge of the procesing of offal a second product was developed, meat and bone meal. This grew rapidly in importance as did a third product, tallow. By 1961 there were thirty-five employees in the processing plant, ten on the pig farm and an office staff in Clonmel some eight miles away.

The growth in electricity demand reflects the growth of the enterprise itself. In 1958 the electricity supply made available through the rural electrification scheme was via a single-phase line erected primarily to supply the requirements of the farmers in the vicinity. In 1960, by adding a third wire, this was converted to three-phase to supply a 50kVA substation at the Ronan pig farm. In 1965 a 200kVA transformer was installed, in 1968 a 400kVA; and by 1974 a 640kVA was required. (As a rough guide to the reader, this last would supply a maximum of about 750 horsepower.) The enterprise and its power demands continued to grow and in 1982 a special heavy power line had to be built direct from the ESB grid station at Cashel to meet the still growing power requirements, which had now approached the 1,000kVA level, and to provide for likely new growth in the future. The plant which had started off in a modest way in 1959 now employs about eighty people in Castleblake alone. A skin and hide business and the accounting department in Clonmel give further employment.

The main product from the Castleblake plant is 'Premier' meat and bone meal, which has a very high protein value and is used widely as a supplement to cereals such as barley to produce a balanced animal ration. The company has also entered the pet-food market both directly, by preparing and canning its own product, and indirectly, by exporting in frozen block form lungs and liver to UK pet-food firms for further processing and marketing under their own brand-names.

The firm is constantly improving and up-dating its production methods to improve quality and reduce energy demands. It has, for example, recently installed modern electro-mechanical plant for removal of water from offal replacing to a great extent the older oil-hungry steam process. It is currently investigating the latest developments in the processing of certain animal constituents for human medication. The existence of the rural electrification networks will ensure that adequate power will be available for any proposed or envisaged expansion in the years to come.

Many more such examples could be given to illustrate the potential commercial and manufacturing talent which existed in rural Ireland awaiting only the advent of electric power and the stimulus of the financial aid, advice and encouragement provided by the official development bodies. In the development of rural industries, electrification has contributed on two fronts — by providing the power for the industries themselves and by making possible a rural living standard which competes successfully with that of the cities for the requisite work force. Over the sixties and seventies a large number of workers gladly returned with their families from the industrial cities of Britain to take up employment with new industries in their home localities — frequently at a reduced level of pay. In 1969, Lucey and Kaldor published an analysis[4] of the impact of industrialisation on two predominantly rural communities in western Ireland, Tubbercurry, Co. Sligo and Scarriff, Co. Clare. The study showed that two-thirds of the Tubbercurry employees and over half of those in Scarriff were rural residents mostly living on farms. Most of them continued to do farm work in their off-time. There had been an increase in population of 231 in Tubbercurry and 318 in Scarriff over what it would have otherwise have been, due to immigration and non-emigration.

The impact of this dispersed industrialisation is particularly evident in counties Galway, Mayo, Kerry and Donegal. In counties Sligo and Clare, apart from the two small centres of Tubbercurry and Scarriff, industry has tended to be more centralised in Sligo town and Shannon industrial estates. Even in these cases, however, a large percentage of the work-force lives in the surrounding rural area where the essential amenities of living are also available with the added bonus of a country environment.

CHAPTER TWENTY

Financing — the Post-Development Phase

PLANNED POST DEVELOPMENT

AT THE END OF 1960 with over a quarter of a million consumers connected the completion of the original development scheme was in sight and the government requested a review of the position and suggestions as to future developments. In March 1961 this review was sent to the government.

The review stated that by the end of 1961/62 all areas qualifying under the current arrangements would have been developed and would be completed by the Board under the agreed subsidy arrangements. Supply would then have been extended to 280,000 premises in 775 areas leaving approximately 100,000 premises unconnected in these areas. This corresponded to the implementation of the scheme originally outlined in the White Paper of 1944. The capital investment in the scheme would then have reached £31½m of which only £6¼m would have been obtained by way of capital subsidy. The overall effect on the Board's finances at the above stage would be an annual burden of the order of £1m.

There would then remain seventeen areas containing about 6,000 premises which would not have qualified for development because of low return. In addition to these there would be approximately 100,000 premises remaining unconnected in the developed areas. The estimated capital cost of connecting all the above would be about £10m. From the Board's experience it was estimated that about 5,000 to 6,000 of these householders would seek to be connected each year for the following 15–20 years at a cost of £500,000 to £600,000 in each year.

However, the magnitude of the current annual deficit on Rural Electrification was such a serious handicap that the Board could not contemplate adding to it by carrying out further development without adequate subsidy. After 'a full and sympathetic consideration of the factors involved' the review concluded that this would mean the provision by the government of the capital free of cost (i.e. 100% subsidy). In the absence of an adequate subsidy the Board could undertake further development only on the basis of setting aside for rural development a portion of any revenue surplus it might earn in a particular year.

THE INTERDEPARTMENTAL COMMITTEE

Having considered both the clamour from householders who had 'missed the bus' in the initial development and the financial position of the Board, the government decided in July 1961 to set up an Interdepartmental Committee to consider the position which would arise on the completion early in 1962 of the current scheme of Rural Electrification and to submit a report and recommendations.

The committee consisted of Dermot O'Riordan (Secretary of Department) and Niall A. O'Brien, Department of Transport and Power, Charles J. Byrne, Department of Finance, Seán MagFhloinn, Roinn na Gaeltachta, Desmond Roche, Department of Local Government, P. J. Brennan, Department of Agriculture, C. A. Barry, Department of Industry and Commerce, Patrick J. Dowling and John B. O'Donoghue, ESB. Patrick Moriarty and Michael Shiel, ESB, attended some of the meetings and Matthew J. Brophy, Department of Transport and Power acted as secretary. The committee submitted its report on 15 January 1962.

The report recommended that the initial development of the seventeen outstanding areas should be completed and that an extra grant of £90,000 should be made to cover the cost of main feeder lines in these areas bringing the total subsidy to £300,000 out of the estimated £500,000 required. This work should be done as a logical completion of the initial development scheme. It was estimated that some 2,500 to 3,000 dwellings out of the 6,000 in these areas could then be connected at standard rates.

It was estimated that there would be some 112,000 unconnected premises on completion of the seventeen outstanding areas. The committee recommended that the 50% subsidy should be continued and connection offered on the basis of a 7.5% minimum return, which was only slightly higher than the 7.3% currently required for post development connections.

The committee recommended the retention of special service charges, as a safeguard against the Board's being compelled to supply even the most remote and uneconomic premises.

It was estimated that under the above conditions supply could be extended to all but 20,000 or so of the remaining 112,000 premises at reasonable charges – from basic fixed charge up to basic plus an ssc of not more than 100% of basic (i.e. twice the basic in all).

For the remaining 20,000 premises the committee considered that electricity supply would be too costly and recommended that a special grant of £10 per premises be made available to these to enable bottled gas installations to be made free of charge.

As the committee saw it, the ultimate position would be as follows:

Electricity		
Number of premises connected at normal fixed charges. (This included 268,000 existing inclusive of about 2,000 in the 17 areas yet to be developed plus 54,000 new connections.)	322,000	(82%)
Number of premises connected with special service charges. (This included 14,000 existing plus 30,600 new connections.)	44,600	(11%)
Total connected to electricity	366,600	(93%)
Bottled Gas		
Total premises with subsidised bottled gas installations	20,000	(5%)
Unconnected		
Total premises not connected for electricity and not qualifying for bottled gas subsidy (i.e. not involving an ssc exceeding 100% of Basic FC). (It was assumed that many if not all of this category would install bottled gas at their own expense.)	7,400	(2%)

A minority report was submitted by the Department of Finance and two reservations were registered by the ESB.

MINORITY REPORT

The substance of the minority report of the Department of Finance was that originally it was never contemplated that every house in the country would receive a supply of electricity under the scheme. The White Paper had assumed a 69% connection. In fact by 31 March 1962, 75% of rural dwellings would have been connected. In the 775 areas already developed some 108,000 houses remained unconnected. Of these 100,000 had refused to take supply when it was offered to them on conditions which would not be improved on under the extension of the scheme proposed in the Report. Of that 100,000 some 50,000 had been quoted standard rates without special service charges. There was in fact no pressing demand for electricity from these areas or from the seventeen undeveloped areas.

The Department considered that the State had more than done its part in assisting rural electrification. While the objective was the provision of a desirable amenity and of a means of increasing agricultural production, in fact on most farms electricity contributed more to comfort than to output. It was essential, now that we had applied to join the Common Market, that a growing proportion of the country's resources should be channelled into productive development.

Neither was the Department in favour of the bottled gas subsidy scheme. The reservations by the ESB were:

The ESB did not consider that rural consumers would accept bottled gas as a permanent and reasonable alternative to electricity. It also considered that, except in the most remote areas, the subsidisation of bottled gas would have an adverse effect on the Board's business.

Secondly, the recommendation that the subsidy should continue at 50% would mean that the special service charges would apply to about half (56,000) of the remaining houses. This would add seriously to the resentment against the Board which these charges engendered. Currently, only about 14,000 consumers out of 280,000 were paying special service charges.

On the other hand, the Board's proposal made in March 1961 to the Minister involving 100% subsidy would, if accepted, permit the connection of virtually all remaining houses at normal rates of charge and would,

A brighter outlook for schoolchildren in the Black Valley with the advent of electricity.

in addition, allow an easement or abolition of the special service charges for 6,000 out of the 14,000 currently affected.

SUBSIDY IS INCREASED

The ESB's arguments obviously made some impression on the government. The Electricity (Supply) (Amendment) Act 1962 which was subsequently passed made provision for the continued subsidy of rural electrification (from this point on mostly post-development work) at the rate of 75% of capital cost up to £75 per house. In return, the Board agreed to extend supply at a minimum return of 4.5% provided the return on the basic fixed charge was not below 2.9%.

As the initial development came to its close, the transition to planned post development (PPD) was effected smoothly. Gangs went back to the developed areas in the order of original development where this was possible, preceded by Area Organisers who carried out a canvass of the unconnected consumers. Of the 10,800 new rural connections in 1963/64, 7,700 were post development consumers. By the end of 1965, these connections were running at the rate of over 10,000 per annum.

However, 1965 brought national financial problems which extended into 1966 and had the effect of decelerating the pace of the PPD scheme. A large adverse balance of payments developed in the first half of 1965 and in July the government introduced a corrective package, which included among other measures a reduction in the public capital programme already fixed for that year. By 1966 the number of new connections had dropped to well under one half that of previous years.

PRIORITY CONSUMERS AND SYSTEM IMPROVEMENTS

At this stage it is necessary to go back some years to follow the build-up of two other demands on the available rural electrification capital resources, 'priority' consumers and 'system improvements' (i.e. increasing the capacity of the networks), both of which had a retarding effect on the progress of the planned post-development scheme.

In the early years of the Rural Electrification Scheme the economy of the country was, if not stagnant, growing very slowly compared with that of other western European countries. In rural areas and in the agricultural sector this was particularly so. Marshall Aid was of great help in the post-war years in launching the large public capital programme (which included rural electrification) aimed at rebuilding and improving the infrastructure. This was also helped by a plentiful supply of labour.

There was also, in the immediate post-war years, a large accumulation of external reserves, which, however, were run down by the mid-fifties.

This rebuilding was not accompanied by any appreciable growth in agricultural productivity (except perhaps in some of the richer eastern and southern areas). Consequently the average rural dweller's standard of living, already low by European standards, fell even further behind. Emigration was rife, especially in the poorer agricultural districts. There was an overdependence on grazing and (as pointed out in the Department of Finance minority report), little inclination or indeed incentive to use the newly available electricity for other than amenity purposes. The result was a slow growth in the use of electricity in the home, in its utilisation for farming activities and in the development of rural based industries. There was not much demand for increasing the capacity of the networks to meet new or growing loads. The ESB rural crews could concentrate their efforts on the planned aspects of the scheme.

EXPANSION IN ECONOMY BRINGS PROBLEMS

Following on the Whitaker report in 1958 and the first programme for economic expansion, a change became evident in rural Ireland. Slowly at first and then more rapidly, the economic pace quickened. More and more use was made of electricity in the home and on the farm; small industries were set up in rural areas; hotels and guest houses opened their doors to the revived tourist industry; new houses were built. All these made new demands on the rural electrification resources. By the mid-sixties a very large proportion of 'rural' capital was being allocated to the reinforcement of networks to cater for increased electricity demand and extending new supply to these consumers, who were considered 'priority' consumers as they formed an important part of the national economy, providing employment and earning valuable foreign currency.

When capital had been more freely available the ESB had been able to cope with the demand from these applicants and also to keep to its programme of providing supply under the PPD scheme. There were tens of thousands of householders who for various reasons had not accepted supply under the initial development but who now were exerting strong political pressure for connection. As available capital became scarce the reinforcement of the networks to keep pace with growth in demand and the provision of new supply to the priority consumers left very little residue of capital for other connections. Again and again throughout the country PPD programmes were postponed. Promises of supply given in good faith were constantly being broken as the pressure for priority work pushed back the connection of these premises.

At the beginning of 1965/66 the ESB's proposed rural capital budget was £2.3m made up as follows: system improvements £340,000, new 'priority' business £500,000, planned post development £1,460,000. As a part of the reduction by the government of the public capital programme, the capital available for rural electrification for the year was reduced to £1.05m. Since the demands for network reinforcement and priority new connections continued perforce at the same rate or indeed slightly higher than in previous years, the full impact of the reduction fell on the post development programme. At the time the axe fell, there were about 3,000 prospective consumers who had their houses wired up, to whom a definite commitment had been made and who now were being pushed back in the programme for possibly another year. Again the government and the ESB were subjected to strong and widespread criticism.

A meeting was held in the Department of Transport and Power attended by the Chairman of the ESB and the Engineer-in-Charge of rural electrification. A further £300,000 in capital expenditure was authorised, enabling the most pressing commitments to be met. Agreement was also reached on the 'priority list' which now included new tenancies by existing consumers, workshops, garages, industries, churches, schools, hotels and guest-houses.

SPECIAL SERVICE CHARGES AGAIN AN ISSUE

The question of high special service charges continued to preoccupy the government. Obviously the political pressures to eliminate these were intense and the government was most anxious to relieve the situation. It was estimated that the authorised capital expenditure of £42m would be reached in August 1968 and a new Bill increasing the legislative limit of rural electrification expenditure was due to be introduced in the Dáil in the new year. This gave an opportunity to review the situation, as any alternative subsidy arrangements which might relieve the position could be incorporated in this Bill and, if agreed, be given immediate implementation.

SPECIAL SERVICE CHARGES REDUCED

The ESB carried out a survey of all unsupplied rural premises and of the degree to which special service charges would apply under the current subsidy arrangements. It found that for 12,000 consumers the level of special charges would not be unreasonable (up to 50% of basic charge), while for 23,500 others they would range from 'high' to 'extremely high'. It developed a package proposal for government approval: if the government, while maintaining the existing 75% capital subsidy, were to extend the cash limits per house from £75 to £150 the ESB could reduce the minimum return required on the capital investment to 4.5% for *all* premises. This would reduce very considerably the level of special service charges, particularly for very remote consumers.

The government refused to consider extending the subsidy limit. However, to the Board's consternation, it was instructed to implement the other half of its proposed package – the reduction of the minimum required return to 4.5% in all cases, even to existing consumers. An accompanying undertaking from the government 'to review the situation at the end of two years' was accepted by the Board with a pinch of salt.

The good news was quickly conveyed to the rural community in a government press release on 20 May 1968. To the ESB, however, the refusal to extend the subsidy limits to balance the concessions now being given was yet another nail in the coffin of the original subsidy concept. The loss on the rural revenue account for the previous year had been £1.7 million and the accumulated loss had now reached over £11 million. The latest directive would greatly increase the cross-subsidy required from the urban electricity consumers. The Board had little option, however, but to accept, however reluctantly, the directive given.

The easing of conditions was like the bursting of a dam. There was a flood of applications for connection. Householders who up to now had considered the terms too stringent, clamoured for supply. As well as those who had held back during the initial development, a new category of premises started to grow in importance, reflecting the improved economic status of the rural-dweller, the newly built house. A new house might be built by a young couple setting up home or it could be a replacement for an existing house. Where the householders had been electricity consumers in their previous residence, or where one of the couple getting married was the son or daughter of an existing consumer, they were regarded as having prior claim to those who had failed to avail of previous opportunities to obtain supply.

Thus, another grouping was added to the already formidable list of 'priority' cases. From the mid-sixties on, these made more and more inroads into the available capital. It was soon realised that unless the capital budget was increased, connections under the PPD scheme, demands for which were also increasing, would have to be drastically reduced. However, the government's capital budgeting procedure, operating as it did on an annual basis did not easily allow budget increases once the estimates for the year had been agreed. On the other hand, the

ESB was expected to respond to demands at short notice, particularly in the case of priority consumers. Furthermore, capital costs of connection had commenced to increase rapidly. This increase in costs coupled with a fixed annual budget meant fewer connections in the non-priority categories. Loud protests came from those householders whose connectioin had to be postponed. The ESB found itself, not for the first time, caught in a pincer squeeze. It was coming under strong political pressure to speed up its connections and was meeting equally strong resistance from the government to spending more than allowed for in the annual budget.

By 1969 the rate of new house building in rural areas was heading for 4,500 per year – all in the 'priority' category. Supply to these and the improvement of supply to existing consumers threatened to absorb almost all the capital allocation for rural electrification. Because of the financial situation, this had been held down to £2 million per annum instead of the £3 million required. The cut of £1 million meant in effect that only £0.2 million would be available for PPD work with a drastically reduced level of connections (only about five hundred instead of three thousand per annum). To make matters worse, this reduction was in the context of a greatly increased demand following the introduction of the more attractive conditions for connection.

In trying to complete the PPD scheme the ESB was aiming at a receding target. The resurgence in the economy in the sixties had extended to the rural areas. A faster growth of real income per capita, reduced emigration, earlier marriages and an expansion in agricultural output, in house building and in tourism all contributed to the demand for more electricity and for connections to new houses, rural industries, commercial undertakings, hotels and guest-houses. It was obvious that unless a completely new approach was developed the PPD scheme would perish from capital starvation.

THE 1971 PACKAGE

In June 1970 James J. Kelly was appointed ESB chief executive. He was a man with little time for pussyfooting or for clawing his way through a web of intangibles, variables and uncertainties. He liked to face problems squarely and seek positive solutions. In one of his first major policy submissions to the Board, he outlined the problem.

Under the initial Rural Electrification Scheme, completed in 1963/64, 792 areas had been supplied with electricity and in these 273,600 premises had been connected at a cost of £32.2 million of which £6.9 million had been provided as subsidy by the government. Up to 31 March 1970, the planned post development scheme initiated in 1961 had been extended to 505 areas. The extra premises connected under this Scheme and under the 'priority' umbrella amounted to about 78,000. The cost of this work and of necessary system improvement work was £14.25 million of which £7.1 million had been provided as subsidy. Thus at 31 March 1970 351,600 rural consumers had been connected at a total cost of £46½ million of which £14 million (or 30%) had been provided as subsidy. The subsidy, of course, was entirely insufficient to avoid an operating loss on the scheme. This was expected to be £2.2 million per annum for the financial year 1969/70. If work continued to proceed at the current rate and on the current basis, it was estimated that by the end of 1974/75 the loss would reach £3.36 million per annum.

Looking to the future, it had been indicated that the maximum subsidy available over the five years 31 March 1970 to 31 March 1975 would be at the rate of £1.8 million per annum or £9 million over the five-year period. The government had also indicated that 31 March 1975 was the date set by it for the termination of rural subsidy. From that time on, new connections would have to be made on the ESB's normal commercial basis, which in the case of rural dwellings could involve substantial capital contributions from the householder concerned.

A change in the whole approach to rural electrification was required if the scheme was ever to reach completion. The constraints resulting from the linking of annual capital expenditure to the annual amount of subsidy agreed by the government were having serious effects.

Completion date was a receding target. The capital available for planned post development was reducing while the number of potential connections was increasing.

It was no longer possible to plan a rural programme.

The national objective of rural supply was not being met but was being pushed aside under current arrangements.

The Board's reputation for service was under severe and justifiable criticism. This criticism was mounting.

There was an inherent injustice to many potential consumers under the current arrangements.

The situation could not be allowed to continue and it was proposed that the Board itself should now take the initiative by setting a firm programme *which did not depend on the current subsidy allocation.* Having set the programme, the Board should then seek discussions with the government to achieve the best subsidy arrangements which the financial situation would allow. The proposal was approved by the Board and welcomed by the government which was of course under severe political pressure to achieve the completion of the scheme as quickly as possible.

From the Board's point of view, it marked the abandonment of the social service approach to the problem with progress dependent on the arbitrary annual allocation of capital. Instead it would now treat it as an electricity supply problem, to be assisted by a contribution from the government up to 1975, but leaving the ESB free to plan it as a single operational exercise. The new approach would accelerate the growing deficit on rural activities, but, as pointed out by the Board's Chairman in a letter to the Minister for Transport and Power (July 1970), the Board found itself in a position where it had little option but to decide on a scheme which would dispose of the problem finally. The proposals did not mean that the ESB believed that a reduced state contribution over the five-year period represented a fair distribution of the burden of cost, but rather that the Board was so anxious to finish the job that it was prepared to shoulder a greater share of the cost.

The Chairman also pointed out that the original philosophy of the Rural Electrification Scheme was that it should involve no loss to the ESB. The concept had long since been forgotten. It was abandoned in the first instance when the government in 1955 withdrew the subsidy, and the loss which had been steadily growing had increased again in recent years by the reduction of the special service charges. Therefore if the Board was to tackle the problem on the lines indicated there were a number of corollaries which should be accepted by the government. These included the ESB's right to retain special service charges, to incur the extra losses involved and recoup them by tariff adjustment and cross-subsidisation. It was also necessary that there would be no hold-up on the necessary tariff adjustments and that there would be an end to the stop-go policies on the capital involved which had so bedevilled the programme in recent years. The total estimated cost of the five-year scheme, allowing for inflation, was £21 million of which system improvements at £12.5 million was the major item. It was estimated that 17,000 PPD consumers and 19,000 'priority' consumers would be connected at final costs of £3.5 million and £5 million respectively.

The government agreed to contribute £10 million as a final subsidy contribution after which it no longer would be responsible for subsidising rural electrification. Due to financial stringency this subsidy would not be paid on the basis of work done as heretofore, but in instalments up to 1977/78.

The main parameters of the new scheme were as follows:

The unconnected houses in those areas in which PPD had been completed would get one further opportunity of availing of supply under the scheme. This would be publicised by advertisements in local papers.

In PPD areas in which work was in progress or in prospect all unconnected householders would be visited by ESB canvassers.

Each area would be deemed 'closed' as work was completed except for certain categories which would continue to be connected on a subsidised basis up to March 1975.[1] Any other applicant for supply in 'closed' areas would be quoted terms based on principles similar to those quoted to non-rural consumers. In these cases, if the quotations were accepted, supply would be extended with the minimum possible delay as was the practice in urban locations.

In this way, the residents of every unconnected private or farming residence in every rural area in the country would have one final opportunity to avail of supply on subsidised terms after which this would no longer be available.

In 1971, the separate rural revenue account was discontinued by agreement with the government in anticipation of the Rural Electrification Scheme ending in 1975. The loss shown in the rural revenue account for that year was £3,254,800 and the accumulated loss to date £19 million.

As so often in the past, events overcame the predictions on which the new programme was based. The prospect of early membership of the EEC brought with it a further quickening of the economic pace in rural Ireland. In 1970 farmers achieved record expansion in output. There was a great swing from milk to beef. 1971 was a good year for cattle but was eclipsed by 1972. In the midsummer of that year, six months before Ireland's official entry into the EEC, there was a huge rise in cattle prices from the low base to which they had hitherto been tied by British cheap food policy. By this time also milk had come back onto the scene as the number of milch cows was expanded. By 1973 agriculture was booming as Ireland anticipated the benefits of EEC membership, wider markets and higher prices.

This boom was directly reflected in the rush of householders to get electricity connection. In contrast to the position at the commencement of the fifties their economic state and social outlook was now such that electricity was regarded as an essential amenity and aid to production. There was also the knowledge that if they missed this last chance of connection under the subsidised scheme, they would have to pay much more for connection later. By the summer of 1974 an updating of the 1971-75 scheme in the light of actual experience indicated that the original estimate of 36,000 consumers to be connected was far short of the mark. It was now estimated that 57,200 would require supply and that the programme would now overspill into 1975/76. This high connection figure would be achieved at the expense of the planned system improvements programme which had fallen very far behind schedule as resources had to be diverted to the now more pressing task of new connections. (The system improvements backlog was to be taken up in the succeeding years and financed completely from the Board's own resources.) Not all new connections were of domestic premises. In 1973/74 for example, 180 applications for supply to new industries in rural locations were processed costing £214,000. None of these industries had been envisaged a couple of years before.

By 1976 it could be said that at last the Rural Electrification Scheme had been completed. The original scheme had been designed to connect 280,000 rural premises at a cost based on pre-war prices of £14 million. In 1976 the Minister for Transport and Power was able to report in the Dáil that 420,000 houses had been connected to the rural electrification networks at a total cost of about £80 million of which £28 million represented State subsidy.

POSTSCRIPT

There was, however, a postscript to the Scheme. In the original White Paper it was accepted that 14% of all rural dwellings, or about 56,000, were so remote as to be outside the scope of any practical electrification scheme. It was also thought at the time that the pressure from these remote dwellings for inclusion would not be very great. Over the years, however, as their isolation lessened through better communications and as their perceived entitlement to a better standard of living grew, the pressures by these householders for inclusion in the scheme became irresistible. Year by year the number of houses excluded was whittled down and by 1976 when the subsidised scheme was scheduled to end, it was a bare 1-2%, the 4,000 to 8,000 (an exact count was difficult) most remote premises in the country.

In 1976 the Minister for Transport and Power introduced the Electricity (Supply) (Amendment) Bill 1976. In his introduction he pointed out that in the previous thirty years some 420,000 houses had been connected to the rural electricity networks at a total cost of about £80 million, of which some £28 million represented state subsidy. This represented some

98% to 99% of all rural houses. There were still anomalies. The Minister instanced the plight of the Black Valley (Co. Kerry) and Ballycroy (Co. Mayo) as being probably the worst. These were isolated communities which, while offered supply under the 1971 to 1975 phase, were not in a position to accept owing to the very high costs involved and very high capital contributions required.

The Bill proposed that householders in these circumstances, i.e. who had refused supply under the final phase of the scheme because of the requirement to pay capital contributions, would be given a further and final opportunity to obtain supply on subsidised terms but, this time, without having to pay a capital contribution. The government in effect would pay the capital contribution. The householder would be liable for the annual special service charge. In taking steps to eliminate this anomaly, the government felt it was going as far as it could reasonably go.

The above circumstances were not confined to the householders in the Black Valley and the householders in Ballycroy. The exact number was not known at the time but a quick ESB estimate had given a figure of 800 to 900 householders. This was not definitive as there were also individual houses in this category here and there throughout the country and the final total might well be more than the ESB estimate.

Following the passage of the Act a detailed survey was carried out and a total of 1,580 premises established as qualifying for subsidy, almost all in the poorer western counties. The total capital cost was estimated as a result of the survey at £3.5 million of which, under the terms of the Act, the ESB would provide £1.6 million and the State £1.9 million.

Even this was not the end of the story. A more prosperous and more progressive Ireland was reflected in the steadily growing consumption of electricity over the 1970s. In 1950 the average annual electricity consumption per rural domestic consumer was 570 units; in 1960 it was 910 units and in 1970 average annual consumption had reached 1,830 units. Up to the 1970's the rural electricity networks as originally constructed had mostly been adequate to cater for the demand. In the cases where reinforcement was necessary due to rapidly growing local demand, it was generally possible to fit this into the construction programme without much difficulty. The general upsurge in consumption over the 1970s as the connection figure approached 100%, coupled with a greater demand per consumer, threw a great strain on much of the network and necessitated the introduction of a programme of large-scale reinforcement. However, the clamour for new connections was so loud and so pressing that the ESB concentrated on the connection of new consumers at the expense of the reinforcement programme. With the achievement of 98-

99% connection by 1976 it was now possible to allocate more attention to the reinforcement (or 'system improvements') programme.

The building of new houses in rural areas went on unabated over the late seventies rising to more than 10,000 per year. Even though the subsidised electrification schemes were now at an end (with the exception of the limited 1976 scheme), the owners of these newly-built houses expected electricity to be made available as soon as the house was finished albeit on non-subsidised terms of supply usually involving a substantial capital contribution by the householder. (An instalment scheme for the payment of such contributions was introduced by the ESB in 1976.) Thus, although the Rural Electrification Scheme as such had come to an end, the period 1976 to 1980 was one of great activity in three different areas of rural electricity supply: system improvements, extension of supply to the houses qualifying under the 1976 legislation and extension of supply on a non-subsidised basis to the 10,000 houses being built each year in rural locations.

The first consumer under the Rural Electrification Scheme had been switched on in the village of Oldtown, Co. Dublin, on 15 January 1947. The objective of the scheme at that time was to connect 280,000 consumers or 69% of the then estimated 402,000 rural premises. By 31 March 1980 the number of electricity consumers in rural locations had reached the figure of 468,000 and, owing to the building of new houses, was growing at a rate something in excess of 10,000 per annum. Because of inevitable overlapping, it is not possible to give exact figures for the different phases of rural electrification, but in round figures, progress was as follows:

	New Consumers Connected
1st phase 1946-1963	280,000
1st phase post-development 1963-1971	60,000
2nd phase post-development 1971-1976 (including 1,600 consumers under 1976 Act)	62,000
Unsubsidised rural connections to 31 March 1980	66,000
Total at 31 March 1980	468,000

The total capital expenditure in connecting this figure, including network reinforcement as demand grew, came to £109,355,000, of which £27,900,000 had been provided by government subsidy.

CHAPTER TWENTY-ONE

A Tale of Two Parishes

EVERY ONE of the eight hundred or so rural parishes has its own story of progress and of the contribution made by rural electrification to its development. To document these stories fully would require many books. I have therefore taken the story of just one small area as a microcosm of the larger canvas — two parishes in west Donegal, an area endowed with striking natural beauty but with few natural resources, to which rural electrification was extended in 1952.

As one travels westward from the town of Donegal, past the busy fishing port of Killybegs, to the two westernmost parishes of Kilcar and Glencolmcille, the land gets poorer and more mountainous, and the landscape progressively more beautiful. The heavily indented and rocky coastline includes some of the highest sea cliffs in Europe, while inland are spectacular vistas of mountain, moorland and glen. The glens between the barren mountains and parts of the sea coast where the land is somewhat less hostile, have been settled for thousands of years: the court cairns in the shadow of Slieve League and in the glen itself testify to the presence of human beings some five thousand years ago.[1] In more recent centuries the indigenous inhabitants of the glens and sea coasts were supplemented by the dispossessed tenants of more fertile lands to the east resulting in a very large population in a very restricted area with few natural resources except turf for the fire, rough grazing for sheep on the rainswept mountain pastures and the plentiful fish in the surrounding waters. It was on these resources that the large population eked out subsistence, spinning, dyeing, weaving and knitting the wool, and fishing from small boats close to the shore. Families were large and, while the harsh environment bred a hardy and resourceful people, it was not possible to feed many mouths from the tiny holdings of poor land or from the proceeds of the cottage industries.

Emigration was inevitable — often by the most active and adventurous young people who frequently rose to positions of great eminence in their newly adopted lands. If the emigrants prospered, a very significant addition was made to the resources of the family remaining home by way of what were officially known as 'emigrants remittances', the periodic cheque from the son or daughter abroad. This was indeed the only sizeable source of cash income for many of the poorest families in the region.

The high rate of emigration in the first half of the present century gave great cause for concern. In the 1950s a young curate in the parish of Glencolmcille, Fr James McDyer, became so distressed at the draining of the lifeblood from the parish that he initiated strenuous efforts to develop local projects to provide employment and so help to keep the young people at home. His efforts and achievements in the parish itself have passed into history. He awakened in the nation as a whole a consciousness of the danger of losing a valuable part of its cultural heritage if parishes such as Glencolmcille were allowed to die.

Fr McDyer saw in rural electrification a means of halting and even reversing the outward flow of people. Electricity could help in the development of local industries, including the tourist industry; it could brighten up the homes of the people and dispel the mental and physical gloom; it could ensure an attractive social life and through television open up new ideas and concepts.

When in 1950 the rural power lines started to snake westwards from Donegal Town a new hope was thereby awakened for the future of these west Donegal parishes. Through rock, bog and swamp to Mountcharles and Dunkineely, Killybegs and into the parish of Kilcar the lines were driven, extending the new power to hundreds of rural and village dwellers. By the summer of 1952 the parish of Kilcar had electricity and the Glen River boundary of the parish of Glencolmcille was reached. Across the river to the village of Carrick and three miles south to the tiny harbour of Teelin the lines stretched so that by Christmas the inhabitants on the east side of Glencolmcille parish were enjoying the benefits of electric power. Now, however, there was a full stop in the westward thrust. Beyond Carrick was a barren five-mile tract of moorland and beyond this the glen itself, the onetime abode of the saint from whom the parish drew its name. Even on the basis of the liberal terms made possible by the government subsidy, the number of householders in the glen itself and in the neighbouring glens who were prepared at this stage to avail of supply was too small. The construction crews were withdrawn to bring supply to other areas in the county and it appeared that it would be many years before they would return.

Fr James McDyer with colleague at Glencolmcille.

Fr McDyer recalled:

> we ran into fairly determined opposition by the few. These were mainly ageing householders who had no families . . . However it would be difficult to blame them. Their attitude sprang from their traditionally disenfranchised existence and their isolation from involvement in progress. After all, were they not to fear another alien force from outside. They had good reason to suspect such forces, and progress. . . . However I was determined on this occasion that I would not be thwarted because there could be no worthwhile progress without electricity.
>
> I asked the Electricity Supply Board to send down one of their best men to address a general meeting. They did better. They sent down Colm Browne who was a native of the district. The meeting in the community hall was packed, which was a good omen. Colm really excelled himself and I was sure I could feel the vibrations of assent from the audience.[2]

Fr McDyer then set his parish council to recovering the situation. Evening after evening they called into the small houses explaining, persuading and cajoling the householders into signing the required application forms so as to secure the numbers and revenue required to ensure electrification. The benefits to the individual and to the locality as a whole were teased out over and over again. So were the objections that electricity would be too dear and too dangerous in a thatched cottage. The economic state of the majority of householders was indeed such as to give them pause before committing themselves to paying a fixed charge of from £6 to £8 a year. The cash income of the majority was small, for some perhaps the sale of a few sheep, cattle or wool; for others payment for piecework knitting or weaving or the occasional money order from the son or daughter abroad. While the people were self-sufficient to a great extent, certain commodities had to be purchased and there was no money to spare for unnecessary luxuries. What if they found it difficult or impossible to meet these new bills? Many preferred to forego the promised benefits rather than risk the humiliation of disconnection at some future date.

However, the dedication and persuasive powers of Fr McDyer and his small band achieved success. Sufficient signatures were obtained and in December 1954 an historic and significant little procession paraded in period costume down the village street of Glencolmcille. In the front was a child carrying a rush light, followed in turn by other children carrying other means of illumination – a tallow candle, a wax candle, a single wick oil lamp, a double wick lamp, an incandescent lamp and finally an electric bulb. At the local church prayers were said; a blessing was given, a switch was thrown and electrification of the parish of Glencolmcille was at last complete.

It had not been easy. The engineer's report on the area reflects some of the construction difficulties, especially through the many bogs traversed by the lines.

> Of 900 poles erected, over 700 had to be fitted with stays to ensure stability. One pole, 42 ft in height . . . kept sinking for a fortnight and eventually decided to say put at a depth of 10½ ft . . . Servicing also was very difficult, the houses being long, low and thatched, with very poor walls and no chimneys . . . of the total number which had originally signed application forms, nineteen dropped out. Of these five houses had become vacant, eight householders were now building new homes and in two cases family bereavement had enforced a change of view. However, sixty-eight new consumers were gained and the annual fixed charge revenue rose from £1,072 to £1,370. Capital costs rose more steeply however due to construction difficulties.

Finally the report speaks very highly of Rev Fr McDyer, CC, and his parish council for the considerable help they gave in keeping down backsliding.

One of the problems recalled by Fr McDyer was the difficulty in recruiting local labour for the scheme because of the fear of the small farmers of losing the dole.

> I wrote a trenchant letter to the late Bill Norton who was Minister for Social Welfare in the government of the time. . . . On the following Monday upwards of thirty extra workers were directed by the Labour Exchange to report for work. It was well for me that my action was never known because my popularity would have plummeted.[3]

In 1983, some twenty-nine years later, a visitor to the parishes of Kilcar and Glencolmcille who had not seen the district since those early fifties would have noted tremendous changes.

Local industries in the two parishes and the development of the fishing port of Killybegs ensured that employment was available to any young person who wished to remain and settle down. No longer was the emigrant ship the inevitable fate of many young people. Many thatched cabins had been replaced by modern bungalows with electricity, running water and modern sanitation. New houses were constantly being built as young people, now assured of a living locally, got married and set up their homes. People from outside the parish, from outside the county and indeed from outside the country itself moved in, attracted by the beauty of the locality now supplemented by the availability of modern living amenities.

The growth of local industries has been promising. Tourism in this remote and beautiful region has been given a boost by the hotel, guest-house and 'bed and breakfast' accommodation made possible by the ready availability of electricity and water. Television, which is installed in almost every house, has banished any sense of remoteness as has the wide ownership of motor cars made possible by the wages now coming into most houses.

THE TWEED INDUSTRY

Donegal tweed from the two parishes has long been famous for its quality, texture and colour. The industry had been under the auspices of the Congested Districts Board as far back as the 1890s. From 1928 it was organised and encouraged by Gaeltarra Éireann, and up to the late 1960s a successful business based on the handwoven cloth was operated. Skilled handweavers were employed at the factory at Kilcar, at marts in various locations and in their own homes. The coming of rural electrification in the early 1950s improved conditions immensely, particularly by providing good lighting for the work during the long winter evenings. However, with the advent of free trade in the early 1970s, it was obvious that new and higher levels of productivity, quality control and design would have to be achieved to place Donegal tweed in the high fashion category on the export market at keenly competitive prices. It was necessary to modernise and mechanise and introduce sophisticated design and marketing techniques. The challenge was taken up with vigour and the result in a modern industrial complex in Kilcar under the aegis of Údarás na Gaeltachta (the successor to Gaeltarra Éireann) employing over seventy people in the tweed factory and another 130 in the ancillary dyeing and spinning plants with an estimated sales figure of £1½m for the year 1980. While most of the workers are drawn from the two parishes, it has been necessary to recruit from as far away as Ardara to fill some vacancies.

FISH PROCESSING

Fishing has long been traditional in the locality, as has fish processing. Filleting and smoking of fish were encouraged by the Congested Districts Board at Teelin and Killybegs from the end of the last century and by Sir Burton Conygham at Rutland, Burtonport, as early as 1780. Killybegs has long been a major fishing port and a large number of ancillary industries has grown up since the coming of the rural electrification networks ensured power for expansion. The growth of fishing in Killybegs has provided many jobs for the people of the neighbouring parishes of Kilcar and Glencolmcille. More recently, however, smaller but no less successful fish-processing industries have been developed in the parishes themselves.

At the tiny port of Teelin, from which small fishing boats operate, a local man, Mr Jack Gallagher, operates a small but very modern enterprise which purchases, processes and markets the catch from the small local fishing boats as well as a proportion from the larger boats operating from Killybegs. Starting with a very small single-phase electricity demand from the local rural electrification network the requirements of the enterprise grew rapidly. By the late 1970s, electric power requirements had risen to over 50hp involving numerous three-phase motors. This demand was met by installing a converter which delivered three-phase supply from the single-phase branch. By 1981, however, demand had risen to about 150hp with blast freezers, cold rooms, smoking plant and electronically controlled vacuum-packing machinery. The original single-phase branch line erected in 1952 to supply the modest domestic needs of the village had finally to be converted to three-phase to meet the demands.

The plant, which between processing staff and fishermen ensured employment – part-time and full-time – for about fifty local people, is equipped to process and produce fish, mostly herring, mackerel and cod, in any form the market requires – fresh, frozen, wet salted, dry-salted, smoked, brined (roll mops) – and ship them in barrels, boxes or individual vacuum packs to markets in Britain, the Continent of Europe and the West Indies. Mr Gallagher told the author of a visit from a buyer for the famous Harrods of London Food Halls who, having tasted a smoked herring in a Paris restaurant, tracked the fish to its source – the tiny processing plant in Teelin – with a view to securing supplies. Unfortunately for Harrods all available supplies had already been contracted for, mostly to French buyers!

Six miles up the Glen River in the townland of Meenaneary on the boundary between the two parishes is the fish processing plant of Earagail Éisc Teo., which formerly was the vegetable packing factory of the Errigal Co-operative Society, one of Fr McDyer's early experiments. In mid-1981 the plant, under the direction of Mr Jarlath Morris and still in the course of expansion, already employed thirty-five local people and has targeted for a total of eighty permanent jobs plus a further thirty seasonal (August to November) jobs.

As in the case of the Teelin plant, the Meenaneary enterprise covers a wide variety of fish processing. Mackerel, which has become the biggest volume landing, is the most promising product. It is generally marketed in frozen form – 'round frozen', frozen fillets and 'headed and gutted' –

for markets in France, Germany and Africa. Herring, on the other hand, is generally salt-cured or smoked and marketed as roll mops and salad fillets to Germany and Holland. Crab, which in early days was frequently thrown overboard when caught, is now a big money earner in the form of whole crab, crab meat, fresh pasteurised vacuum packed bodies and crab claws. Markets are in Sweden (two-thirds of output) and France (one-third). Cod, which generally has a short season (February and March), is usually processed by salting. Good markets exist in Mediterranean and West Indian countries for this salted cold-water fish to replenish salt lost by perspiration.

Unlike most fish-processing plants, that at Meenaneary is not sited at a fishing harbour. It is fifteen miles from Killybegs and six from Teelin. However, Jarlath Morris maintained that the fact that the staff, almost all local including many small landowners, had such a proprietorial interest in the business and such local pride that the productivity of the plant and the quality of the product more than compensated for the extra cost of haulage.

A feature of both the fish-processing plants is the large cold stores where the product can be held for long periods to take advantage of market trends. The growth of the Meenaneary plant can be measured by the growth of the freezing and cold-store requirements

1971 — 45hp blast freezer plus 10hp cold store installed
1975 — 80hp blast freezer plus 20hp cold store added
1981 — Three 100hp blast freezers plus 40/50hp cold store (with capacity of 800 tons) added.

The total requirements of over 500hp are supplied from a rural three-phase line originally erected in 1952 as a single-phase branch to service the then modest electricity demands of the domestic consumers in the locality.

The picture thus emerging from a survey of progress in the two most westerly parishes in County Donegal is that the extension of rural electrification in the 1950s heralded the transformation of a whole rural society of about 2,500 souls. From being a backwater the area has developed into a well-balanced forward-looking community with excellent housing and all the necessary services for comfortable living in a pleasant rural environment. Local manufacturing and processing industries supplement the traditional sheep rearing and fishing to ensure a high level of employment within the parishes themselves. No industry is yet so big as to isolate the workers from their rural background. They continue to live where their roots are instead of disappearing into the anonymity of a big city, thereby enriching not only their own lives, but the life of the community as a whole.

Epilogue

OUR CHRONICLE OF THE PROGRESS of the Rural Electrification Scheme in the preceding chapters has led us from the late forties to the early eighties. This period of little over a generation has seen an immense change in Irish rural society. It would be appropriate in the closing pages of our story to review this change and the contribution made by rural electrification. Better housing, water on tap and modern aids made possible by electricity have greatly enhanced living standards in rural homes, which are no longer so isolated. A more informed and progressive society now obtains as a result of widespread secondary education, the social and cultural activities of the voluntary organisations and the better communications available through telephone, television and radio. (One Area Organiser estimated that in many of the rural areas he canvassed in the late forties only about one family in six had a radio, battery operated and consequently used very sparingly.) The rural community, which now more than ever includes many non-farming families, has developed into a strong and articulate element playing a full part in the life of the nation.

Out on the farm, especially where a new and more enlightened generation has succeeded in wresting control, there is a new and more business-like approach. Electricity, now installed on practically every holding, is playing its part in the application of modern technology. On many farms it is helping to improve quality and output and hold down production costs in the context of a steadily declining work-force. Improved marketing techniques are being employed to develop existing markets and open up new outlets for increased production. Much, however, remains to be done. Irish agriculture is still the most under-developed and undercapitalised in the EEC. Much potentially productive land still stagnates under the control of elderly tradition-bound owners. One authority[1] has estimated that, with proper development, agriculture could provide an extra 75,000 jobs over a fifteen-year period by reducing the drift from the land, increasing input and increasing the volume of goods for processing.

Rural industries are helping to compensate for the declining employment levels in agriculture. Even in the case of large industries centred in the towns the contribution of rural electrification has been considerable. Much of the labour force for these is drawn from the surrounding rural hinterland where the amenities of living now available, combined with a rural environment permit an attractive life style which in turn discourages permanent migration to the towns. It could be claimed that many foreign industrialists have been attracted to set up in Ireland because their executives and work forces could live in pleasant rural surroundings while still enjoying most of the amenities of modern living.

After a long period of decline the rural population is now rising steadily. This reversal is particularly welcome in the west, where the population had continued to drop even through the sixties when the rest of the country was experiencing a revival in growth. Between 1971 and 1979 the population of Connacht increased by 7.1% while that of the three Ulster counties in the Republic rose by 9.1%. In County Donegal the increase was 12.5%.

Utopia is of course far from being reached: at the time of writing there are immense problems, economic and social, looming. The progress of the people of rural Ireland however, in the period under review, has been by any standards remarkable. The contribution to this progress by the Rural Electrification Scheme has been a major one and has fully justified the faith of its architects and that of the founders of the Shannon Scheme in the infant years of the Irish Free State.

The old and the new. *Above*: Milking in the early twentieth century; *below*: A modern electrically powered rotating milking parlour.

APPENDIX ONE

The White Paper

REPORT ON RURAL ELECTRIFICATION

IN PREPARING THE REPORT, Thomas McLaughlin and his team of assistants, P. J. Dowling and A. J. McManus, drew freely on the proceedings on various World Power Conferences, the reports and publications issued by the Rural Electrification Administration of the United States, The Hydro Electric Commission of Ontario, and the British Electrical Development Association.

It first set out the dimensions of the problem.

Of the 221 cities and towns totalling 1.15 millions in population, 193 had ESB supply, 26 were supplied by local undertakings and 2, with a total population of 1,077, were without supply.

Villages of 200–500 inhabitants comprised 215, of which 104 had ESB supply, 28 had supply from local undertakings and 83 (total population of 23,690) were without supply.

There were 386 villages of under 200 inhabitants (total population 47,600) of which 80 (total population 11,200), had ESB supply; 13 (total population 1,700) had supply from local undertakings and 293 (total population 34,700) were without supply.

Finally, there was a scattered rural population of 1.7 million persons of which only 18,000 enjoyed the benefits of public electricity supply.

Part I of the body of the document examined the design, construction and likely costs of a practical rural electrification scheme. It surveyed the existing distribution network and visualised the rural supply system as a natural extension of this. However, it saw that growth would have to be planned on a complete area basis rather than as a 'ribbon' form of development. The line design provided for was such as to permit of the lines being strengthened in a simple and economic manner as the load grew.

Four trial areas were selected to represent the three main categories of farming, milk production, tillage and cattle grazing. These areas, which were (i) near Thurles, Co. Tipperary, (ii) around the village of Glanworth, Co. Cork,

(iii) between Ferns and Enniscorthy, Co. Wexford and (iv) around Ardee, Co. Louth, were also selected to represent areas where the farms were predominantly of large, medium and small size.

Complete detailed engineering designs, working plans and unit costs for the various operations were worked out using a cheaper, simpler and lighter design than for networks designed to serve areas of concentrated population. The total cost of supplying the estimated 402,750 rural premises was then worked out and came to a first approximation of £17 million at pre-war prices, or an average of £42 per premises. Finally, this part of the report emphasised the magnitude of the proposed scheme. On the basis of the average of 5.37 dwellings per mile of supply line given by the models, this would entail 75,000 miles of line. The highest rate of construction hitherto reached at the peak of the Shannon Scheme by the contractors was 650 miles in one year and the highest reached by the ESB itself in the subsequent years was 380 miles per annum. This highlighted the necessity for a completely new type of construction organisation if the rural scheme was to be completed within a reasonable time.

Part II of the report consisted of a review of rural electrification development in other countries. In many European countries the farming community lived as a general rule in large clusters of farmhouses or villages with a consequent high density of dwellings in the area to be served with electricity. Supply to these was, from a technical and economic viewpoint, the same problem as the supply to rural villages and small towns, a problem which had already been solved in Ireland. The Irish farmers, however, generally lived in well-scattered locations, a situation which might best be compared to the United States or Canada. Progress in rural electrification in the U.S. had been impressive. In California it was claimed that 84% of farms had electricity supply and New England, New Jersey, Massachusetts and Connecticut had achieved figures of 93%, 83% and 83% respectively. However, it was emphasised that the U.S. was such a vast area and conditions so divergent in the various states that any discussion on rural electrification as applied to the country as a whole could be of little value.

It was in Canada, in the organisation of electricity supply to rural consumers in the province of Ontario, that an almost exact parallel to the problem under investigation was found. This development was therefore examined in detail and reported on at length. The salient features of the Ontario development were as follows:

The farmers lived in scattered dwellings as in Ireland.

The supply authority was a statutory Board akin to the ESB. Members of the Board were appointed by the government and finances provided by way of advances from the State.

Supply was given from a national electricity network.

The State electricity board organised, built and managed the local rural network and sold direct to the consumer. Development by way of supply to communal rural groups had been tried and abandoned.

In order to bring the burden of the capital expenditure within reasonable economic limits, it was found necessary for the government to subsidise development to the extent of 50% of the capital cost.

In no other country or province studied was there such a close parallel with conditions in Ireland. Furthermore, in no other case was the problem of subsidy dealt with so clearly and cleanly. A large number of subsidy systems and other means of giving financial assistance operating in other countries had been examined but all were in some way over-complicated or, to quote one of the authors, 'messy'.

When the Irish report was being prepared it was not possible, because of the war, to visit and examine systems in other countries. There was almost complete dependence on published reports and other such literature, and in this regard the well-produced Annual Reports of the Hydro Electric Commission of Ontario were of the greatest assistance.

Part III of the report dealt with organisation, management and service. Organisation was approached from two main considerations, technical and administrative.

Technical considerations required that supply must be extended in the first instance from the existing 10,000 volt system. The focal points from which the rural lines would radiate would therefore be the 38,000 volt/10,000 volt transformer stations. This would be done by means of 10kV, three-phase 'backbone' lines. Thus the country would be divided up into a number of large rural supply areas. Each of these would form a unit in itself, but the networks within it would be built up by a gradual process, construction necessarily extending outwards from the existing 10kV system. Within these large rural supply areas, any smaller areas could be taken for construction but it was sensibly enjoined that areas closest to the existing 10kV system must be given priority in development.

The administrative aspect involved consideration as to what form the rural organisation should take. Should administration be undertaken by co-operative associations of rural dwellers (as in Sweden and in many parts of the U.S.) or by the Electricity Supply Board itself? The report gave examples from Sweden of the pitfalls involved in getting small co-operative supply organisations to keep proper books and introduce proper rates of charge. It pointed out that in Ontario where rural networks were developed in the manner and on the national scale contemplated for Ireland, the sale of electricity to communal groups was tried and abandoned in favour of direct sale to individual consumers by the State electricity board. On the other hand the report demonstrated how neatly and economically the administration of the rural scheme in Ireland, if carried out by the ESB would mesh in with the existing 'District' and 'Area' organisation.

Finally, Part III took up the question as to how best the electricity could be used to improve farming productivity. It emphasised the great value placed on rural electrification internationally as a means of improving social conditions in rural areas, lessening the burden of drudgery on farmers and their families, brightening their lives and reducing the great gap existing between the amenities of city and country.

Part IV of the report dealt with the financial aspect of extending the electricity network into the rural areas. Up to now, every extension of supply by the ESB was subject to rigid commercial evaluation and was not undertaken unless it produced an adequate financial return. This would be patently impossible in the case of widespread rural electrification. This part of the report examined the capital involved, the annual costs arising from the capital expenditure and the possible methods of charging the consumer in order to recover these costs. It came down on the side of having a uniform basis of charge for supply to all rural areas. It also accepted that this charge should be in two parts: a 'fixed' charge which was based on the annual costs of the rural distribution network and a 'unit' charge for the actual energy consumed based on the cost of generation and transmission. This 'two part' method of charging was to be one of the most misunderstood and hotly challenged aspects of the scheme.

Having examined the many options on which the fixed charge could be based – valuation of holding, acreage of arable land, the size of the electrical installation among others – and the problems that each system would give rise to, the report finally recommended the floor area of the dwelling-house and out-offices as the basis for charge, very much on the lines of the 1936 Schedule of Rates for Small Villages and Isolated Rural Consumers. While, like all the others, it was not free of anomalies, it had some relation to the quantity of electricity likely to be used and also, in some measure, to the householder's relative ability to carry the financial burden.

The report then went on to consider the scale of the fixed annual charges per consumer on the principle that on the average for all areas, the sum of the fixed annual charges in a rural supply area should not be less than the fixed annual costs for the area. These costs, which included interest on the borrowed capital, sinking fund for repayment, depreciation, operation and maintenance of networks, had been estimated to amount to approximately 12% of the capital cost of the network for the area.

As a first approach to the problem of fixing a scale for the tariff, an estimate was made of the results of applying the scale in the existing rural tariff to the four small-scale trial areas for which plans and estimates of capital costs had been worked out. It was concluded that with the scale of fixed charges in this tariff, the necessary return of 12% on the capital investment would hardly be achieved in practice from the sum of the fixed charges and that it would be unwise to calculate on a better return than 9.7%.

In the examination it became apparent that even a return of this order was dependent on the exclusion of individual premises where the capital cost of connection was excessive in relation to its fixed charge revenue. It was calculated that a limit of 14 times the annual fixed charge (or a 'criterion ratio' of 14), would allow 86% of the dwellings in the trial areas to be connected and give a return of about 10.8%. However, it was thought probable that some 20% of these would choose not to take supply leaving 80% of 86% (or 69%) of the total dwellings in the areas to be connected. The cost per dwelling of connecting these would be higher than if the whole 86% took supply while the average revenue per dwelling need not necessarily be any better. Calculations showed that the total fixed charges desirable might prove to give only 9.7% of the capital cost and therefore to enable this 69% to be connected to the criterion ratio would have to be raised to 15.6 (i.e. all premises costing 15.6 times their annual fixed charge or less would be included).

This 9.7% return on capital expenditure was of course short of the estimated 12% required to break even, but it was considered the best compromise. Raising the fixed charges by about 25% would theoretically bring about a break-even position, but in practice it would probably mean that fewer householders would opt for supply. As it was, under the existing rural tariff, the farming consumer with out-offices would pay on average about 75% more in fixed charges than for a correspondingly sized dwelling in a town or large village. Increasing this differential still further was not considered advisable. Another method of theoretically breaking even would be to lower the criterion ratio, but if this course was followed, a ratio of about 8.5 would be required to break even, which would mean only about a 30% connection — certainly not 'rural electrification' on the extensive scale hoped for.

Having considered these options, the report maintained that the conclusion already arrived at must stand, i.e. that, consistent with a reasonable degree of development in the trial areas, viz. 69% of the dwellings connected, the percentage return on the capital investment would not be greater than 9.7% per annum. Consequently, a deficiency would result (involving some form of subsidisation).

In conclusion, the report referred to the success of rural electrification in other countries and stressed that the development of electricity supply to serve the rural and farming community in Ireland must, as in these other countries, be a gradual process. The problems involved in engineering organisation, management and finance (and their solution) had been clarified by the experience in other countries. Development in Ireland could proceed on much smoother lines and this should lead to speedier progress.

There were, however, many problems and difficulties peculiar to the supply of the rural areas. They had been set out in the report. The primary problems were of a financial nature — the high degree of capital investment called for and the relatively low monetary return that the investment could yield. In other countries this had required State subsidies. Before any progress could be made in the work of organisation and planning of development on a national scale in Ireland, the fundamental question of finance had to be clarified. The extent of the necessary capital called for was set out in the report.

APPENDIX TWO

Definition of 'Rural'

BECAUSE OF THE LARGE AMOUNTS OF MONEY involved in the government subsidy, it was important that from the very outset a clear definition of 'rural' should be agreed between the ESB and the government. As initially agreed, 'rural' included all areas situated outside the following categories.

(i) Places legally defined as County Boroughs, Boroughs, Municipal Boroughs, Urban Districts or Towns.

(ii) All towns and villages of over 250 population.

(iii) All towns and villages listed in the census returns 1936 (i.e. collections of twenty houses and over) already supplied by the ESB.

(iv) Areas occupied by the following categories of consumers:
 (a) consumers on the fringes of networks supplying areas such as (i), (ii) or (iii) above which might be connected LT[1] to such networks;
 (b) loads which were situated in the immediate proximity of areas such as (i), (ii) or (iii) above and which might be considered as owing their position to the proximity of such an area;
 (c) all isolated loads over 100kVA maximum demand.

The general approach in the above definitions was that towns and villages of above 250 population were not intrinsically rural. Similarly, isolated loads of over 100kVA were considered as industrial and did not come within the definition of 'rural' electrification.

By the spring of 1949, however, following the rapid take-off of rural electrification, it had become obvious that many towns and villages of over 250 population could find themselves in the position of seeing ESB supply in the rural areas surrounding them while still being without electricity themselves or dependent on a usually inadequate local supply. Pressure mounted therefore to have them included in the scheme. The inclusion of such centres of population would in nearly every case improve the economies of supply to the adjacent rural areas. Such inclusion, on the other hand, would have only a marginal effect on the finances of the scheme as a whole as the population and fixed charge revenue (of the order of £30,000 per annum) in the villages and towns in question was small in comparison with the population and expected fixed charge revenue (of the order of £1.5 million) covered by the whole rural scheme.

In early summer of that year, agreement was reached with the government on a re-definition in which 'rural' areas were all locations other than the following.

(i) Towns and villages listed in the census returns (i.e. collections of twenty houses and over) which were already supplied by the ESB.

(ii) The following towns not supplied by the ESB
Boyle, Co. Roscommon — Population 2,093
Buncrana, Co. Donegal — Population 2,295.

(iii) Areas occupied by the following categories of consumers:
(a) consumers on the fringes of networks supplying areas under (i) and (ii) who might be connected at low tension to such networks;
(b) loads which were situated in the immediate proximity of areas under (i) and (ii) and which might be considered as owing their position to the proximity of such an area, e.g. institutions, housing schemes.

The principal effect of the alteration in definition was to include within the scope of the scheme sixty-two towns and villages not then supplied by the ESB. These fell mainly into the 250-500 population group with forty-seven villages in this group. There were eleven in the five hundred to one thousand population group and four in the one to two thousand group.

APPENDIX THREE

The Fluctuating Minimum Return

ONE OF THE MOST PUZZLING FINANCIAL ASPECTS of the scheme to the lay person was the frequency with which the minimum return required from individual consumers changed. This was the percentage return that the standard annual fixed charge on the premises gave on the capital cost of supply.

The original minimum return of 5.7% has been pitched, in so far as this could be done in the presence of so many uncertainties, so as to achieve the coverage recommended in the White Paper (i.e. supply at standard rates could be offered to 86% of premises). This would remain valid provided only that there was no change in the original relationship between capital costs and fixed charges. A rise in costs without a corresponding rise in the level of standard fixed charges would mean larger numbers of premises failing to reach the 5.7% return required and consequently reduce the number to which supply could be offered at standard rates.

From the very first years of the scheme this problem was highlighted by a succession of rises in costs of material and labour. There was a strong reluctance on the part of the government and indeed of the ESB itself to match each successive rise in costs with a simultaneous increase in fixed charge rates. (Indeed the term 'fixed charge' itself, as related elsewhere, was interpreted by many as meaning that once set it could never be changed). The first increase in fixed charge rates (10%) was not effected until 1956 by which time over £16 million had been invested in the scheme and capital costs had risen to 150% of those obtaining in 1946.

In order to preserve the originally planned coverage and thus the spirit of the scheme, each substantial rise in capital costs was, therefore, followed by a reduction in the minimum return required. This of course had a depressing effect on the overall economics of the scheme, but it was hoped that in the long run suitable adjustments in the capital subsidy would be obtained in compensation. Table 9 shows the variations in the minimum return required from 1946 to 1970. It will be noted that upward revisions in fixed charge levels (e.g. 13 September 1956 and 22 November 1960) allowed a corresponding raising of the minimum return, but in general the trend is downwards as fixed charge increases failed to keep up with the escalation in capital costs. The downward trend in the minimum return resulted, as a matter of course, in a corresponding downward trend in the return from the scheme as a whole leading to a heavy annual 'loss' on the scheme which reached £3.2m in the year 1970/71, the last year in which separate 'rural' accounts appeared in the ESB Annual Report. In effect, of course, this 'loss' represented the amount of cross-subsidy from the ESB's urban consumers.

TABLE 9
Minimum Return Required

		Areas under Construction		*Post-Development*		*Remarks*
		Farm & domestic	*others*	*Farm & domestic*	*others*	Criterion ration of 14 referred to in White Paper. When Scheme commenced costs had increased 50% & FC by 20%. $14 \times \frac{150}{120} = 17.5$ new CR = 5.7% giving expected return of 7.76
1946	Original scheme	5.7%	7.76%	—	—	Householders returning less than 5.7% (or 17½ to 1) not accepted
Dec. 1947	Original scheme	5.7%	7.6%	7.76%	7.76%	Above now accepted at 7.76% of return
15.11.51	Rise in costs	5.0%	7.76%	5.0% *	7.76%	To preserve coverage *PD return relaxed
12.9.52	Rise in costs	4.2%	7.76%	4.2%	7.76%	Ditto
2.4.54	Directors raise PD return	4.2%	7.76%	6.0%	7.76%	
13.9.56	Rates of Charge revision	4.6%	8.5%	6.6%	8.5%	To preserve coverage
13.10.56	Directors' decision	4.6%	8.5%	9.0% *	9.0%	Capital Contribution to bring up to 9%; no SSC allowed
11.2.58	Pressure from government, Director's decision	4.6%	8.5%	6.6% *	8.5%	Capital Contribution abolished
25.2.58	Rise in capital costs	4.0%	8.5%	6.6%	8.5%	To preserve coverage

26.4.60	Directors' decision	4.25%	8.5%	6.6%	8.5%	
22.11.60	ROC revision	4.70%	9.0%	7.30%	9.00%	To preserve coverage
15.6.62	Increased subsidy	4.70%	9.0%	5.00%	7.40%	
19.10.62	Increase in costs	4.40%	8.50%	4.70%*	7.00%	*With sliding scale
25.6.63	Increase in costs	4.20%	8.10%	4.50%*	6.70%	*With sliding scale

(Last of original areas completed in 1965 — Minimum Return now only applies to PD)

		Farm & domestic	*others*	
21.5.68	Directive from government	4.5%	6.7%	
15.11.68	ROC revision	4.9%	7.2%	To preserve coverage
23.1.70	ROC revision	5.2%	7.7%	Ditto

CRITERION RATIO AND MINIMUM RETURN

It was estimated in the White Paper that the annual fixed charge revenue would have to reach 12% of the gross capital cost for the scheme to break even. It was also held that it would be unwise to calculate on a better overall return than 9.7% by applying the existing 'rural' fixed charge tariff and even this depended on the exclusion of premises where the ratio of capital cost to annual fixed charge revenue was excessive (p. 36 of White Paper). It was estimated that a limit of fourteen times the annual fixed charge (or a criterion ratio of fourteen) would allow 86% of dwellings in trial areas to be offered connection and give a return of 10.8%. However, it was probable that 20% would not take supply, leaving 80% of 86% (or 69%) to be connected. All the above calculations were on the basis of pre-war costs.

When the scheme eventually commenced costs had increased by 50% while the increase in the fixed charge tariffs permitted by the government was only 20%. Therefore, in order to achieve the original coverage, the criterion ratio had to be increased to 17.5 (i.e. $14 \times \frac{150}{120}$), corresponding to a minimum return of 5.7%. On the same basis, the expected original overall return of 9.7% was reduced to 7.76% (i.e. $9.7\% \times \frac{120}{150}$).

APPENDIX FOUR

Fixed and Unit Charges

WHILE THE GOVERNMENT HAD STIPULATED in its original approval of the Report on Rural Electrification in August 1943 that current rural rates were to apply subject to any war increases generally applicable, certain changes on the then existing rural tariff were proposed by the ESB in January 1946 and approved. While keeping the total fixed charge in an average area constant, the out-office fixed charge was reduced by about 33%, the loss in revenue being made good by a 12.8% increase in the fixed charge for dwellings. The increase in fixed charge caused by the inclusion of the out-offices was spread more evenly (as a percentage) between the different dwelling-size groups.

It was also proposed that the scale of fixed charges set out in the report, which were based on 1939 level of capital costs, be increased by 50% to bring them to the 1946 level of costs, but the government would not agree to an increase of more than 20% (letter from Minister, 2 July 1946).

With regard to the unit charges for consumers on the two-part tariff the Engineer-in-Charge REO in a submission to the Board dated 11 July 1946 wrote as follows.

> The unit charges (proposed) are not the same in the existing rural tariff, the published report and the report of 30 January 1946.
> The difference is shown below:

	Size of Block per 2 months	*Price per Unit*
Existing Rural Tariff 1939 plus 40%	60 units	1.75d
	300 units	1.05d
	over 360 units	0.70d
Published Report plus 40%	100 units	2.8d
	over 100 units	1.05d
Tariff proposed in Report of 30th January 1946	80 units	2.5d
	over 80 units	1.0d

> It will be remembered that there was no special rate for Motive Power for farmyard purposes in the published Report tariff or the proposed tariff. This was the reason for increasing both the size and price of the first block of units. In view of the Minister's letter requesting 'that the existing rural tariff rates be made available to

rural consumers', it is recommended that the proposed rates of 2.5d and 1d as set out above be adopted as roughly equivalent. Further, instead of offering the existing special water-heating rate of 0.7d and 0.46d, it is recommended that an 0.75d rate be offered on the proposed rate for all units over, say, 360 units per period.

The proposed unit rate would then be as follows.

For first 80 units (per two-monthly period)	2.5d per unit
For next 280 units	1.0d per unit
All units over 360	0.75d per unit

This was adopted and incorporated into the 1946 'Rural Rates of Charge' leaflet. In June 1951, due to an escalation in the price of coal (British coal at £4 per ton was difficult to obtain and American coal at £8 per ton had to be substituted), 0.3d. per unit was added to all unit charges.

APPENDIX FIVE

Things Chiefly Technical

THE EXISTING SYSTEM

AT THE COMMENCEMENT OF THE SCHEME there was already a fairly widespread electricity network in the State supplying 250,000 urban consumers. A small 110kV main transmission grid linked the main cities of Dublin, Cork, Limerick and Waterford, with extensions to Dundalk and Carrick-on-Shannon. A secondary transmission system of 38kV lines joined these and the other main load centres in a more widespread pattern.

As the 38kV sub-transmission system was already widespread and could be further extended at a reasonable cost, it was decided that there would be no necessity to adopt a higher voltage than the already standard 10kV for rural distribution. Where long distances or high loads caused feeding problems, these could be overcome by providing extra 38kV/10kV supply points from the existing 38kV grid or on extensions of it. By standardising on 10kV as the distribution voltage, it would also be possible to incorporate many of the existing 10kV 'urban' feeders into the rural supply system. In the event, it was found possible, in the case of some very long 10kV feeders where excessive voltage drop was the problem (e.g. in west Cork, west Kerry, Connemara, west Mayo and west Donegal), to delay investment in new feeding points by using 10kV boosting transformers at points along the line. Where the line loading subsequently caught up with the original capacity, new 38kV/10kV injection points were subsequently provided to reduce losses and provide a more secure feed with the original network acting as standby.

BASIC DECISIONS

As the resources of the State were severely limited, it was vital that the maximum coverage should be achieved for the money available, but without departing from acceptable standards of supply. Two issues of importance had to be resolved at the outset: firstly, should supply normally available be three-phase or single-phase, and secondly, what assumptions should be made regarding the demand to be catered for? On these decisions would depend to a great extent the ultimate cost and success of the scheme.

THREE-PHASE VERSUS SINGLE-PHASE

The big advantage of three-phase supply from the farmer's point of view was that three-phase electric motors were simpler, cheaper and more easily obtained, especially in the larger sizes, than single-phase. In the farmhouse, of course three-phase supply had no advantages over single-phase. Many Continental farmers and most British farmers had three-phase supply for motive power and the more progressive Irish farmers were well aware of this fact. However, the average British farm was far larger and had a higher potential electricity demand, while on the Continent the farmhouses and farmyards were often concentrated into sizeable villages requiring three-phase supply in any case (as indeed did similarly sized villages in Ireland).

In Ireland, for historical and other reasons, most farms were small and farmers almost invariably lived on their holdings, which resulted in a very dispersed pattern of load. Furthermore, at the time, the general level of farming activity and standard of living was very low. It could therefore be expected that in the initial years the average demand and consequently the revenue forthcoming from this dispersed pattern would also be low. An average annual consumption of eight hundred to one thousand units per consumer was assumed for the purposes of the original design. The problem was to devise a system that would meet adequately the expected initial demand and require the minimum initial capital investment, but would be capable of expansion easily and at a reasonable cost to meet growing demand.

The solution was the 'dual three-phase and single-phase, phase/phase' distribution system.[1] In this, a three-phase three-wire 10kV line ran through the centre of each area of about 25 square miles. This joined two 38kV stations, could be fed from either source as required and thus formed a secure supply 'backbone'. Branching out from this, single-phase 10kV spur lines connected between two phases, penetrated into all parts of the area, feeding at intervals small single-phase 10kv/LT pole-mounting transformers serving groups of houses or even individual houses.

In most cases these single-phase lines were simply convertible to three-phase by the addition of an extra conductor, while the transformers could easily be replaced by larger sizes as the load grew. The length of a spur or branch could be anything from a few hundred metres to 20km with thirty to forty transformers in some extremely sparsely populated districts. The maximum load in these last was unlikely to exceed 200-250kVA in the initial years. In the case of large groups, the LT lines were so designed that they could easily be converted to 10kV. This would enable a group of houses initially fed by one transformer to be 'split' later into two or more transformer groups at the minimum cost to meet growth in load.

The decision turned out to be a sound one. Initial investment was kept at a minimum and subsequent system reinforcement could be paced to keep up with growing demand and growing revenue. As far as motive power was concerned single-phase motors of up to 3hp could normally be supplied from the single-phase system. They were easily available and while they were somewhat more expensive than their three-phase counterparts, the price difference was only a small fraction of the saving in the supply network costs. 3hp motors were adequate, and indeed still are, for almost every job on the average Irish farm. In the comparatively few cases where larger motors were necessary, up to about 40hp could still be supplied single-phase although these motors were comparatively dear and normally required some local strengthening of the supply arrangements. For requirements above this, three phase was normally made available under a special contract. In recent years, static and rotary converters have made it possible to supply comparatively large three-phase loads, up to 100hp from the single phase system at a fraction of the cost of converting the lines to three phase. The cost saving is particularly high where three-phase supply is required for reasons other than size of load, (e.g. many specialised electrical machines with a comparatively low power demand are available only in three phase form). In one example, the cost of line and transformer conversion would have been in the region of £8,000, whereas the problem was solved by the installation of a converter at a cost of £400. (See Appendix 6 — Three-phase versus single-phase.)

LOAD ASSUMPTIONS

Here again it was important to steer a course between keeping the capital cost to a minimum and meeting adequately the power demands of the consumers. Studies were made on the development of electricity consumption in rural areas of other countries; agricultural experts were consulted, as were numerous potential consumers. As a result, for the purpose of arriving at the design load, an empirical relationship was developed between the size of house and out-offices, as expressed by the Fixed Charge, and the expected electricity demand. This came to 0.4kW of demand per £1 of bimensal fixed charge (or per £6 of annual fixed charge). For very small dwellings a minimum of 0.25kW was taken while even in these cases, if the application form showed that the immediate installation of any substantial appliance was intended, a minimum of 1kW was taken. In the case of a large number of consumers, eight or more, being fed from the one transformer, it was sufficient to add up to the individual loadings to determine the overall load to be catered for. When a smaller number of houses was involved, the aggregate demand was multiplied by a factor ranging from 1.1 for seven houses up to 4.0 for a single house with its own individual transformer thus:

No. of consumers	1	2	3	4	5	6	7	8
Factor	4	2.25	1.6	1.4	1.25	1.15	1.1	1

Thus the design took the fullest advantage of 'diversity' – the fact that many electricity demands are non-coincident – to avoid unnecessary capital investment.

This formula served the ESB well in the initial development stage by permitting the rapid connection of a very large number of consumers at the lowest possible cost while at the same time providing an acceptable standard of electricity supply.

ELECTRICAL DESIGN, 10kV

The backbone of the rural system was to be three-phase 10kV lines running between 38kV/10kV stations. Ideally there would be two such lines connecting adjacent stations and these would be so routed as to feed through the centres of four to six rural areas of 65 to 90 square kilometres (25 to 35 square miles) each.

With the load assumptions described earlier, the 10kV rural distribution network was designed so as to slot in with the minimum disturbances to the existing 38kV and 10kV system. The design provided for the two backbone lines to unite at some distance, about one kilometre, before entering the station compound. This reduced the number of 10kV cubicle outlets necessary at the station. A typical 38kV/10kV station would have one urban outlet for the adjoining town and three rural outlets, each rural line bifurcating to provide two rural backbone lines which again united and terminated at an adjacent station.

Halfway along each backbone line a 'normally open point' was provided, generally an air break switch locked in the open position thus ensuring that each half was under the control of its normal feeding station while allowing alternative feeding in an emergency. With the average loading conditions, it was desirable to have the 38kV/10kV stations not more than 40–50km apart so as to keep the voltage drop to a maximum of 5% at the normally open point 20–25km out. However, as could be expected, local geography and other circumstances frequently made this target impossible to attain.

The conductor originally chosen was 25mm^2 hard-drawn copper for the backbone and main spur lines. The longer single-phase spur lines were also designed to be in 25mm^2 copper and short spurs or branches off the longer spurs feeding three or four transformers could be erected in the cheaper 16mm^2 galvanised steel. Because of the danger of corrosion from the salt-laden atmosphere, steel conductor was not erected within 16km (10 miles) of the sea.

VOLTAGE DROP

The maximum allowable variation from the nominal 220 volts at the consumer's meter was + or − 9%. This gave greater elbow room in the case of spurs at the feeding end of the backbone line (where the 10kV voltage drop was very little) and considerably less for spurs taking off near the normally-open point where the drop on the H.T. might be up to the allowable limit of 5%. The situation was, however, helped considerably by the provision of alternative tappings (off-load), usually + or − 5% on the HT side of the rural transformers. Thus, transformers near the feeding station could be set to provide a ratio of 10,500 volts HT to 220 volts LT while at the remote end the tapping could be 9,500 volts HT to 220 volts LT.

In the feeding stations the provision of on-load tap changing facilities in the 38kV/10kV transformers was of great advantage as they followed the setting up of the output 10kV voltage where the lengths of rural lines were abnormally long, and also the provision of voltage compensation for load increases.

Even in 1983, after an average of twenty seven years of service, the 10kV system as designed was still proving adequate to cater for the greater part of the rural demand. This was achieved by continuously providing extra 38kV/10kV injection points as the load grew to levels far exceeding the original design level. Between 1946 and 1981 the number of these injection points on the 10kV network increased from ninety-seven to about 350.

FIELD DESIGN

The typical rural area contained about five hundred dwellings. The main design instrument was the Ordnance Survey 6 inch map. The initial canvass of the area would have pinpointed and indexed every house on this map and also every house built since the map was made. Other canvass documents indicated for each house its size, fixed charge, whether an 'acceptance' or 'refusal' and, if an 'acceptance', its likely electricity demand.

Working on this information the designer divided the houses to be supplied into groups, each of which could be supplied conveniently at low tension from one transformer. He then determined the transformer size from the load data and selected the transformer locations so that the minimum amount of low tension network was required. The 10kV spur lines feeding the transformers and connecting with the backbone line were then tentatively sketched in so as to minimise the length required. Meeting all these requirements necessitated considerable juggling, but as the designer gained experience, he was able to produce an efficient preliminary design very quickly. He then walked over the ground, frequently finding that he had to alter the route or possibly the transformer location to avoid obstacles not apparent from the map. At this stage he also fixed pole positions, decided spans and pole sizes and generally prepared for the construction stage.

As each spur was tapped off from two phases of the three-phase backbone line, care had to be taken to balance up as far as possible the load on each phase in the area as a whole and to indicate the tappings to be used clearly on the design documents.

CONFIGURATION

For three-phase lines, an equilateral triangular formation was used with phase rotation clockwise looking out from the feeding point. As each half of a backbone line was normally fed from its adjacent 38kV/10kV station, transposition was required at the normally open point to permit back-feeding. At a later stage, horizontal formation for three-phase lines was adopted as this permitted easier conversion from single phase and allowed shorter and cheaper poles to be used for new three-phase lines.

ELECTRICAL DESIGN LOW VOLTAGE

In the case of a large group of houses with an initially low electricity demand, capital was saved by using one transformer where eventually two or more would be required. In this case, provision was made for later conversion of portion of the LT to 10kV by building it initially to the 10kV specification regarding pole size, clearances, conductor size, tension etc. The only difference was the use of LT insulators instead of 10kV. The erection subsequently of a second or third transformer to meet growth of load was very simple and the conversion of the portion of the LT line involved to 10kV was merely a matter of replacing the insulators. In the case of other low voltage lines, feeding houses close together, shorter poles, shorter spans and lower mechanical tensions in the conductor were employed than in the convertible sections. In this latter case the conductors were generally in vertical formation with the neutral on top and fused phase wire at bottom. Until the changeover to aluminium, the copper size normally used in LT groups was 25mm^2 with 50mm^2 used occasionally for heavy loads. When aluminium conductor came into use, 50mm^2 and 25mm^2 sizes with steel core (SCA) were adopted especially where conversion to 10kV was likely. For lower mechanical tensions, similar sizes but without the steel core were used.

HOUSE SERVICES

It was always desirable to have the aerial from the service pole to the house as short as possible to achieve maximum ground clearance, avoid clashing of wires and minimise the pull on the house structure. The conductor used was usually 16mm^2 or 10mm^2 copper or 16mm^2 SCA. The connection at the house end was frequently to a chimney bracket in the case of one-storey houses or to insulators mounted on fishtail stalks set into the house structure where sufficient ground clearance could be achieved. From this point, insulated conductor clipped to the house structure was run to the meter position. Houses with mud walls were still to be found here and there. They required special treatment as their structure could not bear any of the normal fixing arrangements.

MECHANICAL DESIGN

In developing the mechanical design of the network optimum balance was sought between safety, operational demands, continuity of supply, ground clearance under different weather conditions and always the overriding question of cost and availability of materials. A fundamental parameter was the mechanical tension to which the conductor would be pulled as on this would depend the type and strength and hence the cost of the structures used. The experience gained by the ESB in line design, building and operation in the previous twenty years was of immense value in this regard. The quality (99.9% purity) and characteristics of hard-drawn copper had not changed over the war years and the

German (Verband Deutscher Elektrotechniker — VDE) standard[2] which had been applied in the pre-war years was continued.

The selected working tensions were:

12 kg per sq mm for 50mm^2 copper or a total end pull of 600 kg per conductor

19 kg per sq mm for 25mm^2 copper or a total end pull of 475 kg per conductor

24 kg per sq mm for 16mm^2 galvanised steel or a total end pull of 384 kg per conductor.

Using these parameters, sag charts for field use were produced showing the correct sag under various combinations of span and erection temperature to achieve the working tension for each size of conductor.

In the field, 'sagging' was carried out by the chargehand selecting a span of about average length in the middle of the stretch of line and nailing a lath horizontally on each of the adjoining poles at a distance down from the point of suspension corresponding to the correct sag. By sighting between the laths it was simple to determine the required sag had been achieved. During sagging operations the conductor was suspended on free-running pulleys fixed to the cross-arms. The operation involved first pulling up the conductor tightly until it was slightly more taut than the sag required. This was to 'kill' initial elasticity. The tension was then slackened until the conductor dropped to the proper sag when it was transferred to the insulators and bound off. At the maximum loading conditions designed for, a factor of safety of 2 was allowed.

CHANGE FROM COPPER TO STEEL-CORED ALUMINIUM

By early 1947 copper was becoming prohibitively expensive. The price of aluminium on the other hand was remaining stable and the question of substitution was urgently examined. The tensile strength of pure aluminium is very low but a successful combination of a high tensile steel core to take the mechanical tension, overlain with strands of pure aluminium which provided the main electrical path, had been developed and was in use in many undertakings at this time. Canada had been a pioneer in this field and medium low-voltage lines of steel-cored aluminium conductor (or SCA as it had come to be known) had been tried out successfully as early as 1921 by the Hydro Electric Commission of Ontario. Some of these lines had been taken down and examined after twenty-five years and had been found to be in very good condition. The location, however, was far away from the sea and there were widespread doubts about the resistance of the combination of aluminium and steel to electrolytic corrosion when exposed to salt-laden moist winds in coastal areas. Furthermore, at many points the aluminium would still have to be connected to copper terminals or conductor. There would thus have to be bi-metal connectors used at these points and there were again fears of corrosion caused by electrolytic action between dissimilar current-carrying metals in moist conditions.

The general message was, however, that if certain precautions were taken, SCA should prove satisfactory. The decision was taken to change over and with the arrival of the first consignment of SCA in 1948 the use of copper for rural distribution lines was discontinued. For the 10kV lines, two sizes of the new SCA conductor were adopted. To replace the 25mm² copper, No. 1 AWG (aluminium area) SCA comprising six aluminium strands and one steel strand was selected. This had a total cross sectional area of just under 50mm² (actually 49.48mm²) and a conductivity equivalent to 26.67mm² copper. The ultimate tensile strength of the conductor was 1,575 kg. It was known by its international code name of 'Robin'. (All SCA conductor was given code names of birds. Each name stood for a particular size of conductor and was accepted and understood internationally.) At 1950 prices the gross cost per kilometre of a 10kV line erected in Robin was £300 for three phase and £235 for single phase as against £360 and £280 for copper. To replace the 16mm² galvanized steel conductor, No. 4 AWG (aluminium area) SCA was selected. This again had one steel and six aluminium strands. The total area of conductor was just under 25mm² (24.71) and its conductivity equivalent to 13.3mm² copper. The ultimate tensile strength was 830kg. This was used on the ordinary 10kV single phase spur lines and on lightly loaded LT lines. The code name for this conductor was 'Swan'. The gross cost of a 10kV single-phase line erected using this conductor was £175 per kilometre at 1950 prices, as against £210 per kilometre for copper. Finally, for aerial house service, No. 6 AWG, again one steel and six aluminium strands, was used to replace the 16mm² and 10mm² copper conductor. The code name was 'Turkey'.

Sag-charts for use in the field were made out but while there was only one set for copper, which was used for both erection and maintenance work, in the case of SCA there were separate erection and maintenance charts. The erection chart required the conductors to be pulled tightly to produce 5% to 10% (depending on length of span) less sag than the design figure. This allowed for stretch in the conductor which gradually eased out to its permanent sag. The maintenance chart took this stretching into account and was used for regulating SCA lines which had been in service for some time.

Using 10-metre poles the maximum span achievable for a 10kV three-phase 'Robin' line was 90 metres; with 11-metre poles, this could be extended to 98 metres. For single-phase 10kV line in 'Swan' using 11-metre poles, the maximum span was 132 metres if a 168 cm crossarm was used.

SCA VIBRATION PROBLEMS

Aluminium wire is far more brittle than hard-drawn copper and on the earlier Canadian and American SCA lines vibration failures frequently occurred near the insulator binding. This happened on open flat countryside where low winds could set up sympathetic vibrations in the tightly stretched conductor similar to the vibrations in a violin string. The solution was to damp the oscillation by using armour rods at the suspension points. After some years experience it was found that Irish conditions did not lead to this type of vibration failure and the use of armour rods was discontinued. Aluminium tape was used, however, at the insulator bindings to avoid abrasion and chafing of the soft aluminium strands. The binding wire itself was of medium hard aluminium.

ELECTRICAL PROTECTION OF RURAL NETWORKS

10kV three-phase lines leaving 38kV/10kV stations were originally designed to have one of two types of electrical protection, oil circuit breakers with definite inverse time limit overload relay (RIS), which were generally omitted in small stations (under 5 MVA), or high capacity fuse holders with drop-out fuses, normally loaded with 60 amp fuse links. This was not completely satisfactory owing to the length of time required to restore supply even in the case of transient faults. The subsequent introduction of automatic reclosers at main feeding points effected a great improvement in continuity of supply. Two sizes were adopted, set to cut out at 100 to 200 amps according to the loading of the line involved. Reclosers have been found most effective in quickly restoring supply in the case of transient faults caused by lightning or blown debris in a storm while ensuring safe disconnection in the case of a more permanent fault.[3]

Where spurs take off from the backbone line they are protected at the take-off point by 15 amp HT fuses in drop-out fuse holders. Transformers erected directly on the backbone line are similarly fused. There is no further HT protection on the single phase spur lines but on the longer spurs, solid isolating links in standard HT fuse holders are sometimes erected at an intermediate point to facilitate operation and maintenance.

The small number of fuses in series is due to difficulty experienced in achieving selectivity. HT fuses of less than 15 amps have now been dispensed with due to their tendency to blow on very short-time transient faults. Automatic reclosers at the feeding station ensure that in the case of a long duration short-circuit fault, the fuse protecting the faulty section will have blown before the recloser goes through its third cycle, thus isolating this section while maintaining supply to the rest of the network. This keeps area outages to a minimum.

From 1961 onwards, small rural single-phase transformers were fitted with low-cost internal HT fuses. These, by blowing at an early stage in the event of a winding fault, minimise the possibility of a 10kV to LT fault developing which, by superimposing 10kV on the LT supply could lead to a dangerous situation at the house service. They also help to identify a faulty transformer on a long single-phase spur fused only at the take-off point.

EARTH FAULTS

Earth faults are generally caused by a conductor breaking and making contact with the ground or an earthed item of equipment or by a breakdown to earth in the insulation of bushings, insulators etc. Under normal conditions, the capacitances to earth of the three phases are equal and balanced. Under the condition of one phase to earth, the current flowing back to the system via the earthed point is the out of balance capacitance current of the two 'healthy' phases. On the rural distribution system the 10kV network is not earthed and the earth fault currents are too low to blow the line fuse. At the feeding station, however, the neutral point of the 10kV side of the station 38kV/10kV transformer is connected to earth by means of a potential transformer. In the case of an earth fault this detects a potential between neutral and earth and operates an alarm to alert the operator on duty. Supply will be interrupted to the minimum extent while the task of locating the fault goes ahead. If the fault is due to a broken line in a hazardous situation, supply to the section will be immediately disconnected pending repairs. This system ensures that under earth fault conditions there is the minimum interruption in the area as a whole.

LT PROTECTION

There are usually two fuses in series on the LT, one at the transformers and the other, the service fuse, in the consumer's premises. A 33kVA single phase transformer requires a 200 amp fuse while a 15kVA will have a 125 amp. The service fuse at the consumer's premises varies from 30 to 60 amps, depending on the size of the house installation.

EARTHING OF LT

The tank of each transformer is connected to earth at the transformer pole with a maximum value of earth resistance of 20 ohms. For network earthing, the neutral point on the LT side of three-phase transformers is connected to earth, while for single-phase supply the lagging phase is earthed (thereby becoming the neutral). Where possible, this neutral earth is on the first LT pole from the transformer so as to avoid interference with the transformer tank earth. A second neutral earth is provided at the end of the LT run and if there are ten or more LT poles in the group, extra earths are provided. To minimise 'step-voltage' in the ground surrounding the earth pole in the case of a fault, the earth conductor is enclosed in a PVC tube (in pre-PVC days it was wrapped in a vaseline-impregnated tape) from one foot above ground-level to a foot and a half below.

The earth resistance for the neutral of the LT group should not be greater than 100 ohms. In clayey soils the earth electrode is made up of a number of 0.5 inch diameter galvanised steel rods screwed into each other and driven vertically into the ground. Sometimes in the early stages a galvanised steel plate was buried, but in general the steel rods were found to give a lower resistance and involved far less labour. Where rock near the surface prevents the driving of rods a galvanised earth strip is buried in a trench. A cheaper but not so effective method is loosely to loop the earth lead around the pole butt about one foot below ground level. It is usually necessary to instal a number of these to achieve the required protection.

EARTHING AT CONSUMERS PREMISES

Owing to the absence of metallic water pipes it was not possible to use the water mains as an appliance earth as in most towns at the time. At each consumer's premises either an earth rod or buried earth plate was installed to which the frames of electrical appliances were connected. However, the high resistance of this earth connection combined with that of the network neutral to earth inhibited correct fuse-blowing in the event of the appliance frame becoming alive.

Earth leakage trip switches were advocated as protection to ensure circuit isolation in the event of a fault. However, the immediate post-war voltage-operated models were unreliable and unpopular with consumers who objected to paying for this (apparently unnecessary) piece of equipment. To improve the situation, 'neutralising', or using the supply neutral to 'earth' consumers equipment, was introduced in the early fifties and gave positive fuse blowing in the event of an appliance earth fault. Successful operation depended on the continuity of the neutral path back to the transformer and required special measures to ensure this, such as additional neutral earths on the system network itself, compression connectors on the network neutral and duplicate screw connections at the neutral blocks. In recent years the introduction of current-operated earth leakage trip switches has greatly improved the standard of electrical protection available in the consumer's premises. The provision of this last type of protection is now required by the ESB in all new premises.

PERFORMANCE MONITORING AND SYSTEM IMPROVEMENT

The particular characteristics of the rural system are its widespread and dispersed nature with very long lengths of line (almost 100,000km – 60,000 miles – in all) and its huge number of small transformers (about 100,000). Monitoring the performance of this amount of plant and equipment is a formidable task which is now being take over by computer through the development of a distribution management system. This has two main components

Network Digitising □ This involves storing in the computer memory geographical and electrical data on 66,000km (41,000 miles) of 10kV network so that network details in any section are instantly ascertainable.

Transformer Load Management □ By identifying the consumers fed from each of the 100,000 rural transformers and associating this information with the meter reading and revenue billing process, the computer can provide continuous statistical information regarding the electricity demand on each transformer. Regular computer print-outs list every transformer where demand tends to exceed capacity and so ensure timely replacement. This same information, combined with that provided by the network digitising process permits instant assessment of the network feeding conditions in any group and indicates if and where network reinforcement is necessary.

LOOKING BACK

At the time of writing it is thirty-seven years since the first pole was erected at the end of 1946. It is of interest to look back at the performance over this period of the rural network which now supplies almost half a million electricity consumers. In general it would appear that the technical decisions taken were sound. The 10kV three-phase and single-phase lines as designed and constructed are still adequate in the great majority of areas. Load growth has been accommodated in most economic fashion by the provision of extra 38kV/10kV injection points and by the use of 10kV boosters. Lines and transformers have in most cases outlived their 'book' life.

A number of severe storms and blizzards tested the design and construction standards at an early stage. The first big test came in December 1954 when 32,000km (20,000 miles) of line supplying 126,000 consumers had been erected. The blizzard tore across the centre of the country from Brittas and Kells in the east to Clifden in the west. Wet snow consolidated on the lines to a thickness of over four inches accompanied by high velocity winds. Poles, conductor and headgear were broken or twisted, services were pulled out of houses, some chimneys were pulled down. Over 10,000 consumers lost supply. The rural line crews rose to the occasion and all but the most isolated consumers had supply restored within a few days, the majority within hours.

After storms such as this the performance of the networks was analysed. The analysis identified areas of weakness where improvements could be made at reasonable additional cost. With regard to the networks generally, however, it was a question of balance between two requirements, keeping costs to a minimum and achieving the highest possible reliability. A factor in the equation was the speed with which supply could be restored in areas suffering damage from these rare but intense storms, and it was found that increasing the resources allotted to the repair organisation so as to give a faster response would give a better return on investment and a higher overall standard of service than a corresponding investment in the networks. The enthusiasm and commitment of the line crews to restoring supply, often in the most atrocious weather was, and still is, an essential ingredient of such service. Without this, even the best organisation and equipment available would be ineffective.

As the networks have grown, so also has the availability of alternative supply paths to feed the backbone lines. When coupled with the provision of automatic reclosers to restore supply rapidly in the case of transient faults, the result is a high standard of continuity on most of the 10kV network, a standard not far short of that available in the major urban locations. Similarly, the voltage regulation is being continually improved by the provision of extra injection points and other reinforcement. Owing to the large distances and sparse population, some areas on the western seaboard have still problems in continuity.

Lightning storms are also prevalent in these areas. Extra reclosers and lightning arrestors have been installed at key points. As the demand increases and industry hopefully develops in these areas, extra expenditure will be justified in extending the 38kV system to provide loop feeding.

While it is accepted that the general design of the rural networks is satisfactory, with hindsight some of the problems encountered in service might have been avoided. Using Steel-Cored Aluminium conductor near the exposed west coast proved to be an error. Climatic conditions along this coastline resulted in considerably more electrolytic action between the two metals than had been anticipated or indeed hitherto experienced in other countries. The resulting corrosion required the early replacement of SCA line with the well-tried but more expensive copper. The non-tension bi-metal connectors which were used in the initial stages were also subject to heavy corrosion, particularly in moist, salt-laden conditions, resulting in radio interference, heavy voltage drop or total loss of supply. It could also be said with hindsight that erecting such small transformers as 2½kVA and 3kVA was a mistake as in a very short time, with load growth, thousands had to be replaced by larger units at considerable cost to the ESB and inconvenience to consumers. If 5kVA units had been erected originally instead of these, the difference in cost would have been small and the amount of replacement minimised.

In cases such as these it was a question of learning from experience. There were few precedents to guide the Irish pioneers. The verdict today must be that the job was well done; that the Irish rural electricity network is giving a very high standard of service and good return to the nation for the resources invested in its creation.

THE RURAL 'BIBLE'

It was obvious that in a scheme requiring the erection of up to 10,000km (6,000 miles) of electrical distribution lines and the servicing of up to 34,000 houses a year, speed and accuracy of field design and effective construction techniques would be essential. The field engineers would for the most part be very young — energetic and enthusiastic, most probably, but lacking in experience. A comprehensive manual covering the practical aspects of the work and which would complement the technological training of the engineering school was urgently required.

No such manual existed. One of the first tasks of the newly-formed Technical Division under Harold Montgomery and his assistant J. F. Bourke was to assemble all the available practical knowledge on the design and construction of distribution networks in rural locations. From this a field manual was prepared for use by Rural Area Engineers. The authors drew freely not alone on ESB experience and practice but also that of the Rural Electrification Administration of the United States, the Hydro Electric Commission of Ontario and the British Electrical Development Association. The resulting manual, *Design of Rural Networks,* quickly became an indispensable part of every Rural Area Engineer's equipment and earned for itself the title of 'the rural bible'. Several decades later W. F. Roe referred to it as 'one of the keystones of the Rural job, ...it is a matter of interest that this 'Bible' was widely used throughout Europe. I remember getting requests for it from countries as far apart as Iceland and Turkey.'[4]

The manual was produced in loose-leaf form so that it could be kept up to date in technological developments, in the materials situation and in current costs and prices. It was divided into seven main sections.

Preliminary Survey was based on experience in the field by mature surveyors. It gave valuable hints to the beginner in dealing with the various problems and pitfalls which could be encountered.
Electrical Design covered general layout, load and voltage calculations, design economics, conductor and transformer sizes, electrical protection and earthing.
Mechanical Design gave tables of pole sizes, spans and sags for different sizes of conductor and their appropriate tensions enabling quick selection to suit the terrain.
Selection of Materials was one of the largest sections containing as it did tables covering pole and headgear size, spans, conductor spacing, ground clearance, stay size, house servicing, metering etc.
Estimating Costs were broken down under various headings enabling the engineer to give a quick and accurate estimate for any proposed section of work. As material and labour costs escalated rapidly, this portion of the manual required constant revision.
Public Relations included relations with government departments and local authorities. It also covered the ESB's statutory powers and its obligations to the general public and to landowners, wayleaves, compensation for damage, interference with post-office lines, roads, railways, airfields, sports fields, schools etc.
Working Drawings. A comprehensive act of these was provided at the end of the manual and was constantly revised and added to as available materials and technologies changed in the course of the scheme.

As one of the main objects of the manual was to eliminate as far as possible the use of design formulae in the field, there was a comprehensive range of tables and charts whereby the correct pole size, span, spacing etc. could be read off for almost any set of circumstances. These speeded up the work immensely while also giving confidence to the young engineer in the design and layout of the networks. Following the decision in 1947 to change from copper to aluminium for practically all overhead conductor it was necessary to revise radically the design parameters. This involved a major revision in that part of the manual dealing with spans, tensions and clearances and necessitated the preparation of a large number of additional drawings covering the new types of hardwear involved. Even now, thirty-five years after its first appearance, the manual could still be regarded as an up-to-date guide to rural distribution network construction. The performance of the rural lines over the period is a tribute to its effectiveness.

APPENDIX SIX

Three-Phase Versus Single-Phase

WHILE THE NORMAL ELECTRICITY SUPPLY available in urban areas is three-phase alternating current, the stringent economies of rural electrification, particularly with a sparse and dispersed pattern of dwellings, as in Ireland, demanded single-phase supply. As described in Appendix 6 the result was a three-phase backbone and single-phase distribution system which ensured single-phase supply to the maximum number of consumers at the minimum cost. For the first thirty years of the scheme the absence of a generally available three-phase supply was accepted with little question, mainly because the power requirements of the great majority of farmers were confined to pumps, milking machines and small grinders and hammer mills requiring not more than 3hp, which could easily be powered with single-phase motors.

In the case of farms – comparatively few in those early days – with large power requirements, three-phase supply could generally be extended on acceptable terms because of the considerable extra revenue involved. In the early seventies, however, with the economic acceleration of agriculture following Ireland's accession to the EEC, the power requirements of a growing number of farmers began to increase. Larger feed-processing units, larger milking machines, auxiliary equipment for milk-cooling and storage raised the question of the adequacy of single-phase supply. More and more frequently the allegation was made that the lack of three-phase supply was inhibiting expansion of the agricultural sector.

In order to assess the position, an extensive study of the role of three-phase electricity supply in Irish agriculture was carried out in early 1976 by the ESB.[1] The object of the study was to establish in what sectors of agriculture, if any, the extension of three-phase supply would facilitate expansion and contribute towards higher productivity on Irish farms. The study took into consideration the overall economics of extending three-phase supply and of the alternatives (e.g. extra tractor power) in the case of farms with the greatest scope for increasing productivity and profitability. It also set out to derive information on levels of single-phase supply desirable to identify operations to which the application of electricity would be of benefit to agriculture, and sought to involve the agricultural industry to a greater extent in the ESB's forward planning to meet the needs of the industry in the years ahead.

In addition to intensive desk and field research, discussions were held on the project with the Irish Farmers' Association, the Irish Creamery Milk Suppliers'

Association, the Department of Agriculture, the Agricultural Department of University College, Dublin, the Pigs and Bacon Commission, economists, engineers, and agricultural specialists in the research centres of An Foras Talúntais. In addition discussions were held with the staffs of the British National Agricultural Centre and the Electric Farm Centre in Stoneleigh, Warwickshire, and finally with suppliers of equipment and machines. The main findings of the study were as follows.

While lack of three-phase electricity was not inhibiting expansion in the agricultural sector, areas were identified where it could make a positive contribution to increased productivity on the farm.

Of the farmers interviewed who had single-phase supply, not one identified lack of three-phase as a hindrance to expansion or increased productivity.

All farmers would like three-phase supply, for reasons of convenience, provided it was free.

The main requirement for major motive power was for home compounding of stock feed and disposal of slurry.

For pig enterprises over a certain level, the return on investment, taking into account contributions for three-phase electricity supply, as well as costs for storage and milling and mixing equipment, could be very attractive.

In the case of slurry disposal the investigators were not convinced that the power required for the various stages of the process was such as to require a three-phase supply.

Forced ventilation for pig housing and large scale poultry producers could be catered for adequately by single-phase electricity supply.

For the average progressive farmer it was not so much a question of supply being inadequate as that some of the new plant coming on the market required three-phase rather than single-phase supply. This applied particularly in the case of motors. Although single-phase motors of up to 3hp were readily available on the market, above this rating three-phase was the norm. Especially in the case of the larger ratings three-phase motors were cheaper, gave better service and were easier to maintain than single-phase. In addition, some specialised machinery coming into use was equipped with purpose-made three-phase motors even at ratings of less than 3hp. This applied in the case of small rural industries as well as in farming, and also to auxiliary equipment associated with glasshouse heating systems.

While for consumers in proximity to the three-phase backbone line the cost of conversion of their supply to three-phase could be comparatively low, most consumers were not in such a favourable location and quotations of several thousands of pounds for mains conversion were common, effectively putting three-phase outside the means of all but the largest consumers. A breakthrough came with the development of single- to three-phase converters, particularly when in 1977 two Irish manufacturers put on the market models specifically designed for Irish needs. These could be installed in the consumer's premises and could convert the available single-phase supply into three-phase which, while not equalling the quality of the normal three-phase mains supply, was quite adequate for most three-phase equipment. The big advantage was that the cost was generally significantly less than mains conversion, usually less than half.

For some small industries the saving could be much greater. In the case of a rapidly expanding fish processing plant in County Donegal, an investment of £2,000 in 1978 in a converter deferred for several years the necessity to convert a long single-phase spur line at an estimated cost at the time of over £10,000 and thus enabled expansion to take place without an undue financial burden on the enterprise. The cost of a group water supply for Knockbride in County Cavan in 1977 was reduced by £4,000 by the use of a converter, again as an alternative to converting a single-phase mains supply to three-phase. Two principal types of converter were employed — static and rotary. Static converters, which depend mainly on capacitors, are generally more suitable for single motor applications where the converter can be matched exactly to the motor. Rotary converters incorporate an unloaded pilot motor with a single-phase input and three-phase output and are more suitable for multi-motor applications. While the effectiveness of the converters as a substitute for a conventional three-phase supply was undeniable and the economics most attractive, particularly in the case of small loads or as an interim stage in the case of growing loads, it must be said that in the long term and for loads of any appreciable size the most satisfactory solution was the provision of a three-phase mains supply.

APPENDIX SEVEN
Corrosion

WITH PURE ALUMINIUM CONDUCTOR the question of rust, which is such an enemy of ferrous metals, does not arise. In the open, a layer of aluminium oxide quickly forms on the surface, but formation of the layer inhibits further action rendering the conductor virtually corrosion-proof even in moist, salt-laden atmospheres. There are problems, however, when an aluminium conductor carrying an electrical current comes in contact with a dissimilar metal such as steel, brass or copper, in the presence of moisture.

As aluminium has a very low tensile strength it is not on its own suitable for use in overhead lines subject to high mechanical tensions. For a long time this inhibited the substitution of aluminium which was a cheaper metal than copper and not as subject to wild market fluctuations. The development of a conductor with a high-tensile steel core surrounded by aluminium strands – steel-cored aluminium, quickly abbreviated to SCA – opened up the whole field of overhead electricity transmission and distribution to aluminium. It had only about one-half the conductivity of copper so that twice the cross-sectional area were required, but was also only about one-third of the weight so that for similar current-carrying capacity, it was in fact lighter. For rust inhibition, the steel core was galvanised but it was found that in the moist, salt-laden atmospheres such as are found around the coast, corrosion quickly took place. When salt and moisture penetrated between the outer aluminium strands, electrolytic action was set up between the aluminium and the zinc coating of the steel core. The zinc, being the more electro-positive of the two metals was quickly eaten away exposing the steel. At this stage the action reversed and the aluminium, electro-positive with regard to the steel, commenced to disintegrate into a powder. As the current-carrying capacity of the aluminium was reduced, over-heating occurred which accelerated the process until finally all the strands broke. This type of corrosion was confined to a strip about seven miles wide inland from the coast beyond which SCA gave satisfactory service. For this strip some trials were made of the rather expensive all-aluminium-alloy conductor which had 87% of the tensile strength of hard-drawn copper and consequently did not require a steel core. Once again threre were corrosion problems and so in the 1960s it was decided to revert to hard-drawn copper lines for the coastal areas.

The other main corrosion problems occurred where it was necessary to join aluminium conductor to copper such as with house services or transformer or fuse terminals. To effect this, aluminium bi-metal connectors either 'parallel groove' or 'split bolt' type were used. These had copper liners for the copper conductor and again in coastal areas electrolytic action resulted in corrosion and overheating leading to high resistance and failure. Tubular compression type connectors were then introduced, filled with inhibiting grease or graphite, but again similar trouble developed when in time the weather washed the grease off at the exposed end of the connector. In this case, however, it was found that bandaging the finished connection with a vaseline-impregnated tape (Denso tape), so as to seal it off completely from the atmosphere gave reliable results.

APPENDIX EIGHT

Castletownbere — An Adventure in Rural Electrification

(Extract from Final Report on Development of Castletownbere Rural Area by Area Engineer Noel O'Driscoll February 1953)

WHILE BATTLING WITH THE ROCKS in the Schull-Ballydehob fastnesses, blasting our way to our goal we were informed that our next area was Castletownbere. No stick of exploding gelignite could produce more of a stunning effect in the Area Office than when the news reached us. Then we knew that we were being accepted and acknowledged as mountainy men, men of steel and gelignite, capable of shaking still further the serenity of the West Cork mountains whose calm had not been disturbed by the noise of men and clash of steel since O'Sullivan Beara.

In November 1951 I left Ballydehob to visit Castletownbere area, the future scene of our endeavours. Looking at the country between Glengarriff and Castletownbere I wrote off the battle of the Schull area as a skirmish, as I felt that the real battle was here. Here were crags, crevices, canyons, woods, bogs, etc., which defied all exaggeration. W. Trueick, the pegging engineer, was very much depressed at the thought of what lay ahead of him as we climbed up the winding road from Glengarriff to the heights of Loughavaul and beyond again to Coolieragh. However, when we topped the climb at Coolieragh the vista of mountain and sea that met our eyes gave us a temporary respite from our morose reflections.

Here was a scene that is hard to equal anywhere else in Ireland. Ahead of us lay the country of O'Sullivan Beara. Away in the distance lay Beara Island like a sleeping monster resting on the sea, protected on the northern side by the massive bulk of Hungry Hill, and farther west by a ring of mountains whose western slopes dip down into the Atlantic Ocean. Behind us we looked across Bantry Bay at Bantry away in the distance sheltered by the bulk of Whiddy Island. Nearer to us was Glengarriff with its myriad of islets and heavily wooded hinterland, cosy and comfortable looking, secure in the shelter of its encircling mountains.

On a cold November day in the weak wintry sunshine people do not stay long admiring scenery from such a cold vantage point as Coolieragh, and so we continued our journey westwards along by Adrigole, close to the Healy Pass, skirting the foot of Hungry Hill with its silver streak waterfalls and finally we arrived at the capital of the Beara Peninsula, Castletownbere.

From this cursory survey of the countryside over which our backbone line was to be erected, we saw that indeed there were going to be many difficulties. Even at the very start of the line we experienced great difficulty, as the countryside around Glengarriff is very heavily wooded, criss-crossed with rivers and streams and massive boulders rear up their ugly heads to block every feasible path. I called on Head Office in my agony and Mr McEnri arrived and between us we designed a reasonably satisfactory line out of the tangle. Loughavaul was the next pegging worry as here our line had to go right up the steep slope of a mountain and over the top. This was carefully profiled and the difficulty solved by using A-poles on 185 metre spans. This pegging was done in the cold January of 1952 and it is not with happy memories that I recall those cold windswept slopes where we had to use a shield of men on the windward side of the instrument to prevent its blowing over. Hands were cold and handkerchiefs were much in use and many a sheep was routed from his place of refuge behind a rock to give place to our shivering community. The rest of the pegging which Mr Trueick completed was normal by fantastic standards and I am sure Mr Trueick will not recall his mountain sliding with pleasure, as it is not such a sport when one does it on his buttocks. Wayleave difficulties were negligible as no one denied us the doubtful pleasure of planting poles on the virgin rocks. Then, too, as the average size of farm is about 8 acres of land divided into about 16 fields, it would be difficult to put the poles anywhere else but by the fences.

Construction work was exceedingly difficult especially for the first 6 miles of line. Here where no horse or tractor could travel the men had to drag, haul and carry the poles from the roadside up to site on the side of the mountain. It was pathetic to watch the men carrying their pole, now disappearing behind a rock, lost to view for some time, and again reappearing on the slope of another rock farther on, slipping and sliding, and it was not melody that drifted back to the ears of the watcher. It was a great feat to deliver these poles on site without injury to the men but what was still a greater feat was the getting of the compressors up there. This was indeed a marvel of strength and ingenuity, for these compressors were got into positions that even the mountain sheep would declare impossible. Yet neither man nor compressor sustained injury. More tricks were performed with blocks and tackle than would baffle the keenest student of mechanics.

As it was in the end of January 1952 that work commenced, we went about the job in a rather peculiar way. We started at the end of the backbone line and worked back. This was the only sensible way to tackle the job, as any work in the water-logged mountains around Glengarriff would have been impossible at that time of year. The going was tough all the way but there was one very decisive factor in our favour and that was the weather. The local people told us that never in their memory was there such continual fine weather in that part of the country. I will say that the good weather expedited the work considerably and made working conditions reasonably pleasant. While construction work was in progress three other extensions were approved — Glengarriff, Ardgroom and Allihies. The route of the line to Allihies followed an old road which twisted its tortuous way right over the top of the mountain. On the Allihies side of the mountain the slope was steep and so the route had to be chosen very carefully. It was a case of hopping from crag to crag and in all it took only 9 poles to come from the top 900' to sea level. Incidentally, there was no case of 'uplift' thanks to keen eyes. This road over the mountain was never used by mechanical transport before our arrival. It was more like a goat track than an actual road, yet the truck carried the poles up there and the tractor brought up the compressors. Usually the tractor had to help the truck up and this was done by driving a heavy ground pin into the road and attaching blocks. Then the tractor pulled downhill on the blocks and so pulled the truck uphill. I travelled on this road on one occasion in the Supervisor's van . . . I walked it ever after.

There are a considerable number of disused copper workings in this neighbourhood and there seems to be considerable mineral wealth in the mountains all round. Yet no one seems anxious to take the gold out of 'them thar hills'. It was mentioned that there was some move afoot to open up Allihies Mines again, but when and by whom nobody knows. If emigration from that part of the country is to be stopped, some kind of industry will have to be established there, as otherwise the statement of the Land Commission agent when he saw us putting up the poles, 'a waste of good timber', will be true.

Castletownbere was a boom town in the days when there was a garrison on Beara Island and when the Allihies Mines were in production. Fishing was also at its peak at that time and money flowed freely. Now there is nothing left but names in the Labour Exchange. Let us hope that electricity will electrify them into action again.

Construction work ceased in October last and the hills echoed the thunder of our exploding gelignite for the last time. We used 2,950 lbs of gelignite and 3,850 electric detonators. I wonder will that figure be beaten.

There will not be much post-development work, as practically all who were desirous of supply are connected. At the end of October we left Castletownbere, that cosmopolitan town where Irishmen, Frenchmen, Spaniards and Englishmen rub shoulders, and headed eastwards to Cork, 94 miles away.

Notes

CHAPTER ONE

1. On 14 December 1982 a belated tribute was paid by the Institution of Electrical Engineers in the form of a lecture given at its London headquarters entitled 'Nicholas Callan – Neglected Electrical Pioneer'.

CHAPTER TWO

1. Extract from recommendations made by Messrs Siemens Schuckert on the proposed Shannon Scheme and quoted in *REO News*, June 1951.
2. In fact it overran by six months because of a long strike, some very bad weather and the sinking of a ship containing materials.
3. Quoted by P. J. McGilligan in the debate on Rural Electrification in Dáil Éireann, 24 January 1945.

CHAPTER THREE

1. This was raised to 17½ times on the commencement of the scheme.
2. In 1949/50, the amount voted, £325,000, covered the repayment of half of the advances made in the calendar year 1948. In the 1950/51 estimates it was decided by government that the requirements of the 1945 Act would be satisfied by charging to voted monies an annuity designed to repay the subsidy moiety of the advances over the same term of years (50) as the ESB was given to repay its moiety.

CHAPTER FOUR

1. With depreciation and 'other' costs amounting to 7.08% of capital, a return of 8.5% would leave 8.5%–7.08%, or 1.42%, available to meet interest and sinking fund charges of 4.07%. The shortfall therefore would be 2.65%. This is 65% of the total interest and sinking fund charges, requiring a capital subsidy of 65% in order to break even.
2. This practice of lowering the minimum return in the face of rising capital costs was implemented so as to preserve the 'coverage' and it was repeated again and again in the course of the scheme. Ideally, rises in capital costs should have been followed immediately by increases in the rates of fixed charge. For various reasons this was not allowed to happen. Both in timing and in amount, increases in the fixed charge rates tended to lag behind increases in capital costs so that the general trend was to lower the minimum return required so as to preserve the number of premises to which supply could be offered without special service charges or capital contributions. Thus, from a figure of 5.7% in 1946 the minimum return required dropped to 4.0% in 1958. After a brief rise to 4.7% in 1960, it dropped back to 4.2% for the final phases of the original development. Minimum returns required from 'post- development' consumers were. however, almost invariably higher than these. (See Appendix 3.)

3. In *Ireland, Some Problems of a Developing Economy,* edited by A. A. Tait and J. A. Bristow (Dublin, Gill & Macmillan, 1972), there is an analysis by Bristow of this cross-subsidisation within the ESB. His conclusion was that in the year 1968/69, rural prices were lower by over 20% and urban prices higher by almost 9% than they would have been in the absence of cross-subsidisation at the consumption levels of the time.

CHAPTER FIVE

1. '14.3.45 by W. F. R', *REO News*, March 1948.
2. The term REO is used generally throughout the book to denote this organisation. In certain cases, however, the broader title ESB is used where this is considered more appropriate.
3. The accent on youth was not confined to engineers. A random check in a typical area in 1956 gave the ages of the four principal officers as follows: Engineer – 25, Supervisor – 25, Organiser – 25 and Clerk – 23.
4. Apart from the Head Office secretarial staff, REO was an all-male preserve. This of course reflected the culture of the times. To the best of the writer's knowledge not a single female application was received for a job on the rural scheme. In present times it is more than likely that a proportion of the RAEs, AOs, Area Clerks and electricians would be female.

CHAPTER SEVEN

1. In the White Paper the estimated cost of supplying all rural premises in the country, about 400,000, was £17 million at pre-war prices using wooden poles. To supply the 280,000 premises (or 69% of all premises) provided for in the Report would cost £14 million on the same basis. It was estimated that, using wooden poles, the cost in 1946 would have escalated by 50% over pre-war costs to £21 million, while if concrete poles were used the estimated cost would be £28 million.
2. The issue was so heavily oversubscribed that the initial 5*s*. share was dealing at 7*s*. on the day after the issue.
3. By 1982 the company had grown to a group of eight companies operating eleven factories, occupying an area of eighty-three acres in Finglas and employing over 1,200 workers.

CHAPTER EIGHT

1. From an article by Peter Conroy in *REO News*, October 1953.
2. *REO News*, December 1953.
3. *REO News*, November 1954. These were the same gales which had put the pole ship S.S. *Karen* in peril in the Skagerrack (see page 93).
4. *REO News*, May 1955.
5. *REO News*, December 1955.

CHAPTER NINE

1. James Meenan, *The Irish Economy Since 1922*, Liverpool University Press, 1970.
2. R. O'Connor, 'Financial Results on Twenty Farms in Mid-Roscommon in 1945-46', paper read before the Statistical and Social Inquiry Society of Ireland, 29 October 1948.
3. John J. Scully, *Agriculture in the West of Ireland – A Study of the Low Farm Income Problem,* The Stationery Office, Dublin, 1955.
4. Commission on Emigration and other Population Problems 1948-54, *Report* (The Stationery Office, Dublin, 1955).
5. In April 1948 the engineer in charge reported to the Board that there was four times as many applications for electricity supply from the dairying areas as from the non-dairying areas.
6. John Healy, *Nineteen Acres*, Kennys, Galway, 1978.
7. Hugh Brody, *Inishkillane Change and Decline in the West of Ireland*, Allen Lane, The Penguin Press, London, 1973.

CHAPTER TEN

1. Interview with the writer, July 1980.
2. *The Irish Press*, 5 February 1949.
3. It was claimed by traders that the sales of paint in rural areas shot up after the electricity was switched on as the new light showed up grubbiness that hitherto had escaped notice in the pervading gloom.
4. Sometimes a little judicious blackmail spurred the committee to extra effort. In the Blackwater area of County Wexford the search for a suitable office premises was proving fruitless. RAE C. V. Conway rang the District Office in Waterford from the public telephone in the local post office. In a loud, clear voice he told his superiors of the problem and suggested that he might skip Blackwater for the time being and move on to the next area on the list. Quite by accident he chose a time when the post office was crowded with customers. It appeared to be equally coincidental that an hour or so later he was told that a suitable premises had been located by the committee.
5. Stage comedians of the time, particularly in country districts, drew on the scheme for much of their material: 'I have a great job with the ESB – digging holes for poles', or the old chestnut, 'Honour the Light Brigade – Oh what a charge they made!'

CHAPTER FOURTEEN

1. A traditional Irish house Mass dating from the period of anti-Catholic Penal Laws.
2. Michael A. Poole, 'Rural Domestic Electricity Expenditure in the Republic of Ireland', *Irish Geography,* Volume VI, No. 2, Dublin, 1970, The Geographical Society of Ireland.

CHAPTER SIXTEEN

1. Jerome Toner, OP, *Rural Ireland, Some of its Problems*, with comments by Sir Shane Leslie and T. J. Kiernan. Dublin, Clonmore & Reynolds Ltd., London, Burns Oats & Washbourne Ltd., 1955.
2. Stephen Rynne, *Fr John Hayes, Founder of Muintir na Tíre, The People of the Land*, Dublin, Clonmore & Reynolds Ltd., 1960.
3. *REO News*, July 1948.
4. It was about twenty-four inches high and had a central motif depicting An Claidheamh Soluis, the Sword of Light, with two figures representing youth at work, The Sower, and at play, Games. This beautiful carving was the work of John Haugh who at the time was a teacher in Newry vocational school. He had been trained as a woodcarver at Glenstal Abbey and after spending some years carrying out wood carving in Belgium had come to Newry via Marino vocational school, Dublin.
5. Horace Plunkett, Ellice Pilkington and George Russell (AE), *The United Irishwomen, Their Place, Work and Ideals*, Maunsel & Co. Ltd., Dublin, 1911.
6. *Ibid.*
7. For a fuller description of the foundation of the Society and its work, see *The Irish Co-operative Movement, Its History and Development*, by Patrick Bolger, Dublin, IPA, 1977.
8. *Op. Cit.*
9. Muriel Gahan, lecture entitled 'The United Irishwomen', April 1969.
10. *Report of Commission on Emigration and Other Population Problems 1954,* footnote to page 175.
11. Two participants were selected from each county, including the North and South Ridings of Co. Tipperary.

CHAPTER SEVENTEEN

1. Mary Purcell, lecturing on 'The Parish Plan Helps the Woman' in the course of Ladies' Day at Muintir na Tíre Rural Week 1959.
2. John Healy, *Nineteen Acres*, Kennys, Galway, 1978, p. 12.

3. Desmond Roche, 'Rural Water Supply', *Administration*, IPA, Dublin, Winter 1960.

4. Paper, 'Rural Aquafication', read by Fr Joseph Collins at ICA Rural Water Supplies Conference held at An Grianán, 25 October 1960.

5. Desmond Roche, *Op. Cit.*

6. An estimate by the department of Local Government in 1959, quoted by D. Roche (see footnote 3), indicates that the probable percentage of direct service by Local Authority Schemes would be about 50%, leaving half the then unsupplied houses to a greater or lesser extent dependent on their own initiative for a piped supply, helped of course by grants and other forms of assistance.

7. Bernard J. Tighe, 'Rural Water Supply in the Republic of Ireland', *Journal of the Institution of Water Engineers*, August 1966.

CHAPTER EIGHTEEN

1. This however was mainly a substitution: the acreage of oats dropped accordingly.

2. About 95% of this energy bill was for oil used in heating the growing area. Electricity (accounting for about 5%) was used for heating and illumination of growing houses, blowers, circulating pumps, control equipment, lighting, etc.

CHAPTER NINETEEN

1. Commission on Emigration and Other Population Problems 1948-1954, Report, Dublin, The Stationery Office, 1955, paragraph 439.

2. These teams carry out all the initial processing of small industry projects (up to fifty employees) and also provide the after-care service. Priming grants are also available through them from the western fund.

3. In November 1983 Carol Moffett was selected for one of the 'People of the Year' awards, which were presented by an Taoiseach, Dr. Garret FitzGerald.

4. Denis I. F. Lucey and Donald R. Kaldor, *Rural Industrialization – The Impact of Industrialization on Two Rural Communities in Western Ireland*, Geoffrey Chapman, London Dublin Melbourne, 1969.

CHAPTER TWENTY

1. These include industrial consumers, registered hotels and guesthouses on commercial tariff, large intensive farming activities on commercial or industrial tariffs.

CHAPTER TWENTY-ONE

1. 'It may be at first hard to believe that it was the agricultural wealth of these valleys that attracted the first farmers who came to Ireland 5,000 years ago.' From *Glencolmcille, a guide to 5,000 years of history in stone*, by Michael Herity, Elo Press, Dublin, 2nd ed., 1980.

2. James McDyer, *Father McDyer of Glencolumbkille – An Autobiography* Brandon Book Publishers Ltd., Dingle, Co. Kerry, 1982.

3. *Op. cit.*

EPILOGUE

1. Professor Thomas F. Raftery, Head of Department of Agriculture, University College Cork, quoted in *The Irish Times*, 16 February 1983.

APPENDIX TWO

1. 'Low tension', not requiring erection of transformer.

APPENDIX FIVE

1. This system is described in detail in a joint paper read at the 1981 CIRED* conference in Brighton: 'A Distribution System for Rural Zones with Low Consumption Rate' by P. Messager (France), K. Cronshaw (Great Britain), C. Cunningham (Ireland) and M. Deuse (Belgium).

2. This assumes an empirical worst loading condition of $180\sqrt{d}$ grammes per metre length acting vertically downwards at –5°C where d is the diameter of the conductor in millimetres and takes a maximum tension based on the 'maximum permanent static tensile stress' that the conductor can withstand for one year without breaking.

3. The operation sequence of a typical recloser is as follows.

On the occurrence of a short-circuit current exceeding the set figure the switch opens. It recloses after 1½ seconds and if the fault has disappeared, remains closed. If, however, the fault persists, the switch immediately reopens. It repeats this cycle twice and if the fault still persists after the third reclosing, the switch locks out in the open position thereby disconnecting the line involved pending investigation and repair of the fault.

4. Interview with the writer, July 1980.

*Congrès International des Réseaux Electriques de Distribution, organised by IEE Conference Service, Savoy Place, London WC2R OBL.

APPENDIX SIX

1. *The Role of Three-Phase Electricity Supply in Irish Agriculture*, Agricultural Unit, Industrial Development Section, Electricity Supply Board, July 1976.

Index

(T)=Table